The Shoot Apical Meristem
Its Growth and Development

The shoot apex, although tiny and enclosed in the apical bud, forms the whole of the shoot system of plants as well as having a key role in producing leaves and flowers, so an appreciation of how it functions is essential to an understanding of plant growth. In this book the questions of how it grows, and how fast, and the likely cellular processes that are involved in the formation of leaves and flowers are examined at the biochemical, physiological, biophysical, and molecular and genetic levels, in order to try to explain how the shoot apex works.

The shoot apex is characterized as the major site of cell divisions that provides a source of cells for shoot growth. The possible controls on cell division and growth rates are carefully analysed, and the relationship between growth rates and three-dimensional form is explored. The cellular structure, ultrastructure, and physiology are investigated in relation to the processes involved in leaf and floral organ initiation and the generation of form. The newer findings from biophysics and molecular biology are integrated into the evolving picture of the shoot apex.

This is the only book wholly devoted to the growth and physiology of the shoot apex and its key role in the formation of leaves and flowers. It will interest graduate students and researchers in plant development.

Robert F. Lyndon is an honorary fellow of the Institute of Cell and Molecular Biology at the University of Edinburgh and is the author of *Plant Development: The Cellular Basis* (Unwin Hyman, 1990) and of more than 60 papers and articles on plant development, particularly on the shoot apex.

Developmental and Cell Biology Series

SERIES EDITORS

Jonathan B. L. Bard, *Department of Anatomy, Edinburgh University*
Peter W. Barlow, *Long Ashton Research Station, University of Bristol*
Paul B. Green, *Department of Biology, Stanford University*
David L. Kirk, *Department of Biology, Washington University*

The aim of the series is to present relatively short critical accounts of areas of developmental and cell biology where sufficient information has accumulated to allow a considered distillation of the subject. The fine structure of cells, embryology, morphology, physiology, genetics, biochemistry and biophysics are subjects within the scope of the series. The books are intended to interest and instruct advanced undergraduates and graduate students and to make an important contribution to teaching cell and developmental biology. At the same time, they should be of value to biologists who, while not working directly in the area of a particular volume's subject matter, wish to keep abreast of developments relevant to their particular interests.

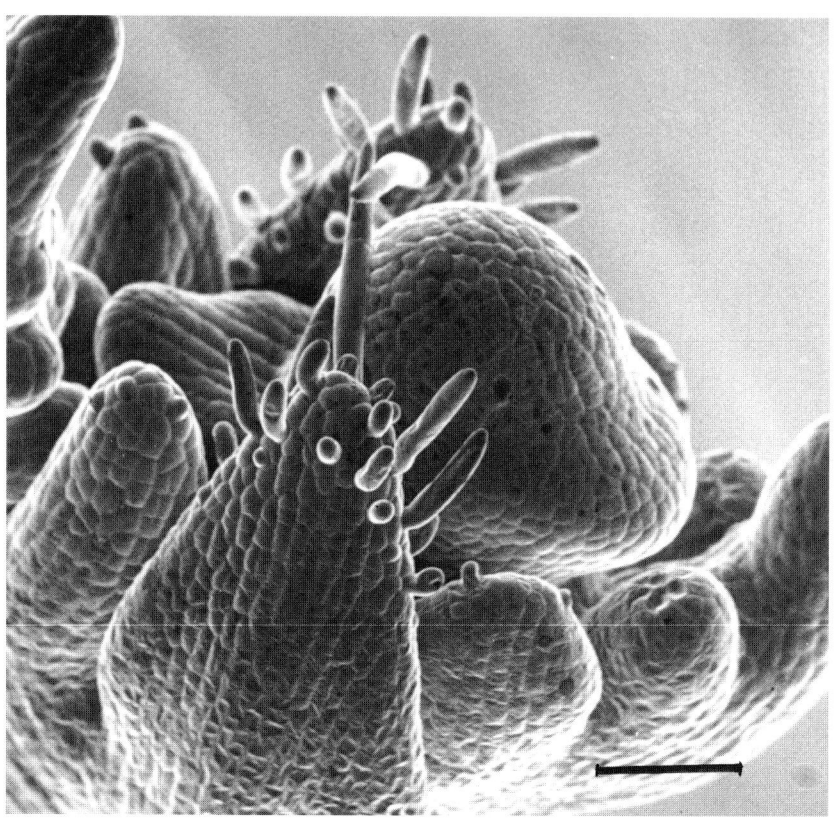

Pea (*Pisum* sativum) shoot apex. The youngest primordium is forming on the right side of the flanks of the apical dome. The previous primordium (on the left) is beginning to show the lobing which is the beginning of leaflet development. The next older primordium shows leaflet development and the stipules (foreground and background) have developed unicellular hairs at their tips. Bar = 50μm.

THE SHOOT APICAL MERISTEM

Its Growth and Development

R. F. LYNDON

CAMBRIDGE
UNIVERSITY PRESS

PUBLISHED BY THE PRESS SYNDICATE OF THE UNIVERSITY OF CAMBRIDGE
The Pitt Building, Trumpington Street, Cambridge CB2 1RP, United Kingdom

CAMBRIDGE UNIVERSITY PRESS
The Edinburgh Building, Cambridge CB2 2RU, UK   http://www.cup.cam.ac.uk
40 West 20th Street, New York, NY 10011-4211, USA   http://www.cup.org
10 Stamford Road, Oakleigh, Melbourne 3166, Australia

First published 1998

Printed in the United States of America

Typeset in Times Roman and Friz Quadrata 10/12pt, in 3B2 [KW]

Library of Congress Cataloging-in-Publication Data
Lyndon, R. F.
The shoot apical meristem : its growth and development / R.F.
Lyndon.
p.   cm. – (Development and cell biology series; v.34)
Includes bibliographical references.
ISBN 0-521-40457-6 (hb)
1. Shoot apical meristems.   I. Title.   II. Series.
QK645.L96   1998
575.4′ 85–dc21   98-10719
                                                        CIP

*A catalog record for this book is available from
the British Library*

ISBN 0 521 40457 6 hardback

Contents

Preface

The shoot apex: what it is and what it does

The meristem of the shoot apex is a remarkable structure: it is about the size of a pinpoint, often only about a tenth of a millimetre in diameter, may consist of fewer than a thousand cells, not apparently very different from each other, yet it gives rise to the whole of the shoot. Shoot meristems are sources of cells that grow and divide to provide virtually all the aerial structures that we see and recognize as plants and that we think of as the vegetation cover of the earth. An individual meristem may perhaps continue to function for hundreds of years, as in old trees such as redwoods and sequoias, or even perhaps for thousands of years as in the oldest bristlecone pines. Although its functioning is highly ordered and cyclic, the shoot meristem can continue to function in the same way when it is isolated from the rest of the plant and cultured.

As well as acting as a source of cells, the shoot apex also forms primordia that develop into the leaves, axillary buds and floral organs. The shoot apex is therefore where the pattern of shoot, leaf and flower development originates. Because the shoot apex works in an obviously repetitive, cyclic, fashion to produce a potentially endless succession of stem modules, it seems equally obvious that there must be feedback controls and cyclic loops in its developmental processes. But at what levels do these operate? Does the cyclic, iterative development of the shoot apex depend on constant alterations in gene transcription or is it a process in which the rules are specified by the genes, but operated by feedback loops at the post-transcriptional level or by metabolic and bio-physical interactions far removed from changes in gene transcription? The view the author inclines to is that form is the result of genes specify-ing the parameters of a system, the operational characteristics of which depend on the physico-chemical properties of its components, rather than each individual component of form and its development being specified by a particular gene. The functioning shoot apex appears to be a system of emergent properties that are not individually specified but depend on the operation of the system as a whole. This point of view is similar to that elaborated much more elegantly by Goodwin (1995, p. 128), who encapsulates it as 'genes don't control; they cooperate in producing variations on generic themes'. However genes act, it is necessary to try to understand how the system works so that we can see what are the processes that need to be explained at the molecular and gene levels.

Despite its obvious importance and our need to understand how it functions so that we can understand shoot development, the shoot apex has been relatively neglected as an experimental object. Unlike the root meristem, which is easily accessible for observation and experimentation, the shoot apex is very small, and is typically enclosed and hidden within the apical bud, so that it can be revealed only after careful dissection.

Because of these difficulties much of what we know about the shoot apex is morphological and has been derived from static, histological preparations. However, the last 30 years or so have seen an increasing use of cellular and biochemical, and more recently genetic and molecular, techniques to explore its structure and especially its functioning. The shoot apex not only undergoes cyclic development, but also shows phase change and developmental switching in a spectacular manner when it flowers. This is making it an increasingly attractive system for the study of the molecular basis of development, especially in conjunction with the use of homeotic mutants, which switch the cells from one mode of development to another.

The theme of this book is essentially an attempt to answer two basic questions: what does the shoot apex do? and how does it do it?

Further reading

This book focuses on the shoot apex and on key references, so many worthwhile papers are not referred to. Much of the earlier experimental work on shoot apices was usefully summarized by Cutter (1965) and Wardlaw (1965). The review of Gifford and Corson (1971) is also a source of references to earlier work. The French school of thought, which stimulated much research but is now largely included within the wider view, was based on Plantefol's theory of phyllotaxis and Buvat's view of the shoot apex as consisting of an 'anneau initial' (initiating ring) which formed the primordia, and a 'méristème d'attente' (waiting meristem – waiting to form flowers), and was the basis of Nougarède's (1967) review of cellular aspects of apical functioning. Plantefol's theory was criticized by Cutter (1959) but strongly supported by Loiseau (1969), who recorded and discussed many fascinating and unusual examples of phyllotaxis. Modern views on phyllotaxis are thoroughly documented and examined by Jean (1994). A useful source of information about the shoot apex, especially in transition to flowering, and flower development, is *The Physiology of Flowering* (Bernier, Kinet and Sachs 1981, Kinet, Sachs and Bernier 1985). Recent valuable books are *Arabidopsis: An Atlas of Morphology and Development* (Bowman 1994), which has many SEM photographs of the *Arabidopsis* shoot apex and the effects of various mutants on its morphology, and *The Development of Flowers* (Greyson 1994), which takes up the story of flower growth from the initiation of the floral organ primordia. The use of molecular biology in conjunction with mutants, to explore the functioning of the shoot apical meristem, is now a fast-moving field. Keeping abreast of it cannot be done adequately in a book, and reference must be made to original papers in journals (among others) such as *Plant Cell*, *Plant Journal*, *Development* and *EMBO Journal*.

Acknowledgements

I am especially grateful to Peter Barlow and Paul Green who read the whole text, to Dennis Francis and Justin Goodrich who read parts of it, and who all made valuable comments on it. However, any errors of omission or commission are solely mine. Acknowledgements are due for permission to reproduce the following figures:

American Journal of Botany (Figs 1.2, 1.3, 1.4, 1.6, 2.6, 6.5); *Canadian Journal of Forest Research* (Figs 3.3, 8.7); Dr R.A.E Tilney-Bassett (Fig. 2.7); Academic Press (Figs 2.3, 3.4, 7.1, 7.2, 7.3, 8.6); *Physiologia Plantarum* (Fig. 9.3); Blackwell Science Ltd (Fig. 8.1); American Society of Plant Physiologists (Fig. 5.5); Gauthier-Villars Editeur (Fig. 5.1); Springer-Verlag (Figs 2.5, 5.4, 7.4, 10.2); International Society of Plant Morphologists (Fig. 9.1); Oxford University Press (Fig. 9.2).

1

A Source of Cells: the Apical Cell

1.1. The apical cell as a source of cells

The apical meristem at the shoot apex generates the cells that divide and grow to form the shoot. Unlike in higher plants, where meristem cells all appear similar (Chapter 2), in lower plants there is a distinctive apical cell which is clearly the source of cells for the formation of the shoot (Gifford 1983). There is usually a well-defined, tetrahedral initial cell at the tip of the shoot which cuts off cells on its three basiscopic faces to produce the rest of the shoot (Fig. 1.1). The cell lineages can be traced in sections so that the shoot can be shown to consist of three sectors of cells, each sector comprising the cells that are the derivatives of one of the basiscopic faces of the apical cell. The three sectors are separated from each other by relatively thick cell walls so that they can be distinguished from each other in sections and in surface views of the apex (Fig. 1.2). The sectors also differ in cell wall cellulose alignment as seen with polarized light (Fig. 1.3) (Lintilhac and Green 1976). Sometimes, instead of being tetra-hedral, the apical cell is pyramidal with four basal faces (Ogura 1972); in some bryophytes (Smith 1955) and ferns (Ogura 1972) the apical cell is wedge-shaped, with only two cutting faces.

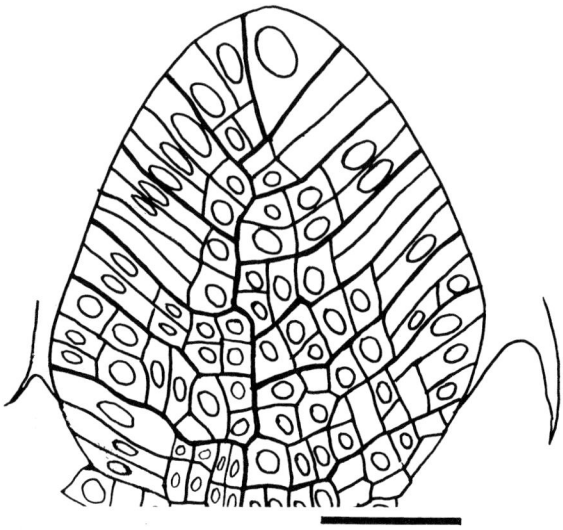

Figure 1.1 *Equisetum* shoot apex (L.S.) with apical cell and the cell pattern resulting from subdivision of merophytes, each the product of a division of the apical cell. The third face of the apical cell is not seen because, while it is in the plane of the section, it is in another section. Bar = 50μm.

Just as the apical face of the higher plant shoot apical meristem does not cut off cells distally (in contrast to the root meristem which does so to form a root cap), the acroscopic face of a shoot apical cell does not cut off cells. The reason for this fundamental difference in behaviour of root and shoot apices is unknown. One face of the shoot apical cell therefore forms the surface of the apical summit and the other three (or two) basal faces are surrounded by the cells of the apex. As the apical cell grows and divides, new cells are cut off in regular clockwise or anticlockwise sequence from each of the three basal faces in strict succession. This can be inferred from the positions of the cell walls as seen in cross section or surface view (Fig. 1.2). The cells derived from each product of division of the apical cell remain recognizable as a group of cells bounded by a thicker wall which corresponds to the wall of the original cell. The derivatives of each cell cut off by division from the apical cell therefore form a group of cells, or unit, having a common origin, termed a merophyte (Fig. 1.2). Each merophyte therefore comprises all the cells derived from a mother cell formed by a single division of the apical cell. The sequence of cell divisions and the cell lineages can be inferred from the positions and orientations of the cell walls, and also their thicknesses, since the older the wall is the thicker it tends to be. If the apical cell cuts off new cells on

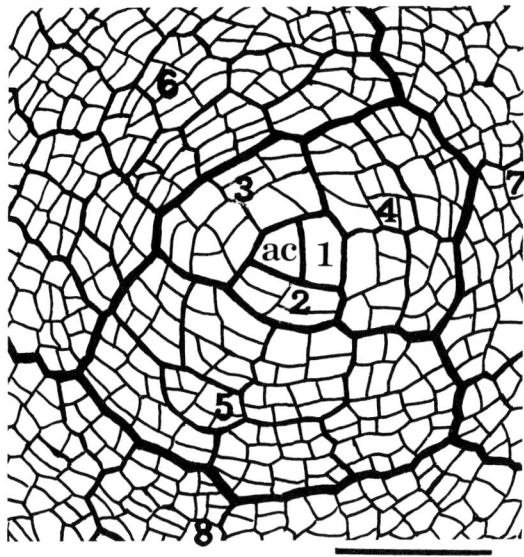

Figure 1.2 Surface view of the shoot apex of the fern *Osmunda*. The apical cell (ac) has cut off cells on three faces to produce the merophytes, outlined by the thick lines, the youngest merophyte being numbered 1, the oldest, 8. Bar = 100μm. From Bierhorst (1977).

three faces, then the shoot consists of three sets of merophytes. Because the site of division rotates round the apical cell, the merophytes form a helix of successively younger structural units up the axis of the shoot.

Where there is a single merophyte contributing to that part of the apex that forms a single leaf, then this means that there is one leaf formed for each division of the apical cell. In this case the rate of division of the apical cell is equal to the rate of leaf formation so that the interval between the formation of successive leaves, the plastochron, is equal to the length of the cell cycle of the apical cell. The rate of division of the apical cell can then be found simply by observing the rate at which leaves are formed.

The shoot apices of plants with apical cells do not appear to grow by tip growth (with growth fastest at the tip), because the doubling times for the apical cell tend to be longer, or the same, but not shorter, than in the more basal derivatives (Chapter 3, Table 3.6). This implies that in the apices of plants with apical cells, as in the higher plants with no single apical cell, there is a gradient of growth rate from a minimum at the tip of the apex to a maximum on the flanks where the leaves form (see Chapter 3).

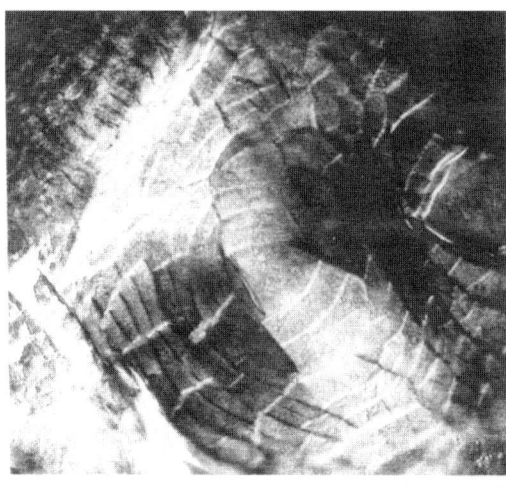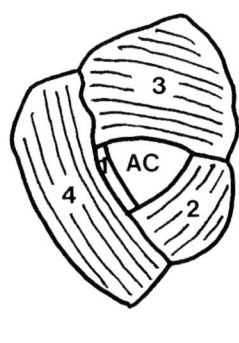

Figure 1.3 Surface of the shoot apex of the fern *Onoclea* as seen in polarized light. Where the main orientation of the cellulose wall microfibrils is NE–SW this shows as a bright area; where the microfibrils are predominantly NW–SE the image is dark. As shown in the interpretive diagram, the main orientation of the wall microfibrils is the same for the derivatives of one face of the apical cell (AC) but differs in each of the three sets of merophytes, the youngest being 1, the oldest 4. Bar = 100μm. From Lintilhac and Green (1976).

The function of the fern apical cell as a source of cells was demonstrated by puncturing it (Wardlaw 1949). Leaf formation in *Dryopteris* then continued only until the existing cells had been used up, implying that the apical cell, once destroyed, was not replaced or regenerated. However, these experiments unavoidably destroyed not only the apical cell but also caused necrosis of the whole of the tip of the shoot apex, so that perhaps potential replacements for the apical cell were also destroyed. In similar experiments with the fern *Osmunda*, the apical cell was destroyed by a precisely positioned 25μm deuteron microbeam (Kuehnert and Miksche 1964), but in this case a new growth centre with cells having the morphology of apical cells developed to one side of the wound and the apex then apparently continued to function normally. Whether a new apical cell can form may depend on the species and the extent of damage done when the apex was wounded.

1.1.1. Do all ferns have a single apical cell?

In the ferns, a single apical cell has not always been easy to identify and so the universal presence of a single apical cell in fern shoots has been questioned. It has been suggested that there is often no single apical initial but a meristem with several initials (McAlpin and White 1974). However, closer scrutiny has shown that in all cases examined an apical cell can indeed be distinguished. A detailed analysis of the surface cell patterns of 150 apices of more than 50 fern species clearly confirmed the classical concept of the universal existence in all ferns of a single apical cell from which all other cells are derived (Bierhorst 1977). The numbers of cells in successive derivative merophytes showed that cell division must have been more frequent in the derivatives of the apical cell, by a factor of about three or four times in some ferns, than in the apical cell itself. This point of view has been supported by Gifford (1983) after weighing all the evidence. However, although study of the segmentation pattern alone can rule out the possibility that the apical cell is dividing faster than its derivatives, it cannot preclude the possibility that it is the derivatives that have become the true initials and that the apical cell and its immediate derivatives form a separate population which no longer contributes to the main plant body (see Section 1.1.2).

The controversy over whether there was an apical cell in ferns (a controversy raised by McAlpin and White (1974), to which Bierhorst (1977) responded so heatedly) arose partly from an inability of some investigators to distinguish the apical cell and partly from a different concept of what constituted an initial cell. Bierhorst's (1977) view was that the term '"initial cell" is used here to refer to a cell of a meristem which serves to perpetuate the meristem. In other words, at least one descendent cell of an initial cell will remain a part of that meristem as long as it exists'. On this extreme view, there will logically be in the meristems of all plants (including higher plants) at least one initial cell even though it may not be a distinctive apical cell. A more useful definition of an initial cell in the lower plants would be one which recognizes that the apical cell is distinctive in structure and position and that it appears to be determined and differentiated as a permanent initial. This is in contrast to higher plants in which no single cell or group of cells seems to be pre-ordained to occupy the terminal position, and the cells that end up being the permanent initials do so by the chance of being in the 'right' position at the right time rather than because of their being differentiated in any discernable way from their derivatives (see Chapter 2).

The clash of concepts in the ferns was also partly because of a recognition that the descriptions of cell lineages, to which Bierhorst (1977) devoted sole attention, are not adequate to describe meristem structure

(White and Turner 1995). Indeed, the two concepts of cell lineage and meristem structure can be considered separately as they are in this book (Chapters 2 and 5). Another aspect of the clash of views was that Bierhorst considered only the surface layer of cells, presumably on the unstated assumption that the derivatives of the apical cell divided to produce the subsurface cells. On the other hand, McAlpin and White were concerned with the three-dimensional functioning of the apex but did not clearly demonstrate whether there was a separate initial or initials for the subepidermal cells. This is a point that Bierhorst simply ignored. However, the lack of chimeras in ferns (Bierhorst 1977) is consistent with there being no separate initials for the underlying cells and with the apical cell being the sole source of cells in the apex in ferns, as it also appears to be from observable cell lineages in the other lower plants.

A further difficulty in accepting a single apical cell as the source of all cells in the fern shoot arose from the interpretation of the undisputed observation (Bierhorst 1977) that in > 500 apices examined only five apical cells were seen in mitosis (a mitotic index of only 1%). Bierhorst concluded that this demonstrated that apical cells did divide; McAlpin and White (1974) cited this same evidence as demonstrating the infrequency of division and relative inactivity of the apical cell. Measurements of the cell cycle in mosses have shown that a mitotic index of about 1% is typical of apical cells and consistent with their being in active division. Although the cell cycle is long, it is of the same sort of length as in the derivative cells (see Chapter 4). Bierhorst's observation of about 1% mitotic index for the apical cell is therefore consistent with its being actively dividing but with a long cell cycle. McAlpin and White, among others, made the mistaken assumption that a low mitotic index necessarily meant a virtually non-dividing cell, which is not so.

1.1.2. Does the apical cell become quiescent or polyploid?

A different controversy arose over the role of the apical cell when it was suggested that in *Equisetum* the apical cell ceased functioning as an initial and progenitor for the rest of the shoot, and that this function was taken over by its derivatives which then became the effective initials. Measurements of the DNA amounts in individual cells in *Equisetum* shoot apices showed that the apical cell and its immediate derivatives could be at a higher ploidy level than the cells below them in the apex

(D'Amato and Avanzi 1968). These lower cells must therefore have been formed from the apical cell before it became polyploid. Subsequently the apical cell would only have been able to form polyploid cells which would constitute a distinct population at the shoot tip. If the shoot continued to grow and form diploid cells then this could only be because the still diploid subapical cells were continuing to divide and would then be acting effectively as initials. It was therefore proposed that the apical cell in *Equisetum* becomes polyploid, produces a few polyploid derivatives, and then becomes quiescent and ceases to contribute to the active growth of the shoot (D'Amato and Avanzi 1968).

The contrary view is that the apical cell is indeed active histogenically and continues to be the source of cells for the formation of the shoot (Gifford 1983, 1985). This conclusion was based on measurements in shoot apices of *Equisetum scirpoides* (Gifford and Kurth 1983). The mitotic index was 3.9% in the apical cells, compared with 3.9% in the subjacent cells and 7.0% in the remaining cells in the apex, and the distribution of the DNA contents of cells between the 2C and 4C values was similar throughout the apex, including the apical cell, with no clear indication of polyploidy. The view that the apical cell is active in histogenesis was also supported by evidence from the aquatic ferns *Ceratopteris* and *Azolla*, which both have slender, finger-like apices, unlike the broad apices of most ferns. In both of these ferns the mitotic index of the apical cell was slightly higher than in the other cells of the apex (Polito 1979, Gifford and Polito 1981). In *Ceratopteris* the cell cycle of the apical cell (69h) was shorter than in the adjacent cells (75h) (Polito 1979), and in *Azolla* it was about the same (27–28h) (Gifford and Polito 1981). In *Ceratopteris* the DNA contents of the apical cell and the other cells of the shoot apex were shown to be similar and to be 2C or 4C, with no evidence of polyploidy even in older plants (Polito 1980). *Nephrolepsis* is a land fern but in the leafless stolons the apex is narrow and conical, is similar in shape to the apex of the water ferns, and the rate of cell division of the apical cell (with a cell cycle of 80h) is almost twice that of its immediate derivatives (with cell cycles of 142h) (Seilhean and Michaux-Ferrière 1985). The rate of division of the apical cell in relation to its derivatives may therefore be a function of the shape of the apex, or, conversely, the shape of the apex is a result of the relative growth rates of the apical cell and its derivatives. In ferns with broad apices, this shape of the apex would be consistent with more rapid growth of the cells below the apical cell in the same sort of way that the primary thickening meristem in higher plants such as palms apparently results from more active growth and cell division in the subapical than in the apical cells.

The differences between the observations of D'Amato's group (D'Amato and Avanzi 1968) and Gifford's (Gifford and Kurth 1983) may be because the former were unintentionally looking at much older, and perhaps senescent, apices. Gifford and Kurth (1983) pointed out that in species such as *Equisetum arvense* (which D'Amato's group used), most of the nodes have been initiated before the shoot breaks through the ground, and therefore at the time of sampling the shoot apex may have become quiescent having already formed the prescribed number of leaf whorls characteristic of the species. It seems that both sets of observations may be correct and may depend on the species and the age of the shoot, as work with other ferns suggests. In *Pteris cretica* apical cells in young meristems had nuclear DNA contents of 2C (40%) or 4C (60%) and, of 50 apical cells examined, none were $> 4C$ (Michaux 1970). However, in 40 adult plants 20% of the apical cells were 2C and 40% 4C, but there were also 35% 8C and 5% 16C, so that 40% of the apical cells were $> 4C$ and therefore polyploid. Polyploidization was concurrent with the accumulation of starch and lipid in the apical cell, perhaps indicating that it had become inactive as an initial. When meristems of *P. cretica* were excised and cultured, the DNA content of the nuclei of the apical cell and its derivatives was never more than 8C, although nuclei in the intact plants were sometimes 16C (Michaux-Ferrière 1973). The percentage of polyploid cells was significantly lower ($p > 0.01$) than in the meristems of the adult, intact plants, again suggesting that polyploidy of the apical cell was more likely as the cell aged. Further studies on *Polypodium* (Michaux-Ferrière 1981a) showed that in the young meristem the cell cycle in the apical cell and axial cells was similar to that in the lateral derivatives, about 62h, but that as the meristem aged the cell cycle in the apical cell and its immediate axial derivatives slowed down to become 144h whereas the lateral cells of the meristem still had a cell cycle of 78h (see Section 4.2.5, Table 4.9). Although this shows that the apical cell divides less actively with age, this in itself does not necessarily imply polyploidization or a lesser importance of the apical cell as the source of shoot cells. However, it is consistent with there being changes in the apical cell as it ages and cautions us against drawing conclusions about apical cell functioning throughout the life of the plant based only on data from young, vigorously growing plants.

The pattern of cell divisions of the apical cell and its derivatives has also been invoked to support the concept of the apical cell being active in histogenesis. It has been claimed that the segmentation pattern of the merophytes was not consistent with the derivative cells themselves being initial cells (Gifford and Kurth 1983, Gifford 1985). However, this argument is unsound because whatever the pattern of segmentation the more distal cells give rise to those more proximal and this occurs at all levels of the meristem, and so the segmentation pattern, whatever it is, cannot be used to support the argument that the apical cell is, or is not, active in histogenesis. As Clowes (1961) argued in the case of the higher plant root meristem, cell patterns encapsulate only the history of previous growth and cannot in themselves tell us about the activity of the cells at the time the apex is sampled. The point made by Gifford (1985) that 'it is difficult to imagine a 2-celled merophyte giving rise to a unicellular merophyte that is in contact with the apical cell . . .' is valid, but it is irrelevant unless one were to suppose that the subapical cells were giving rise to cells distal to them, and there is no reason to believe this. The segmentation pattern is not relevant in considering whether the derivative merophytes become the actual initials. If this happens then the apical group of cells would become an isolated system of cells at the tip of the shoot, as D'Amato and Avanzi (1986) suggest they do.

The conclusion seems clear: in the lower plants and the ferns there is a single apical cell at the tip or centre of the surface of the shoot apex which gives rise to all the rest of the cells in the shoot. There may be some older apices in which the apical cell ceases to function as the source of all cells and this role may then be taken over, perhaps for only a short period before the apex dies, by the immediate derivatives of the apical cell and which lie just below it.

1.2.   Structure and functioning of the apical cell

The apical cell may be easily distinguished by its size, often being much bigger than the other cells at the shoot tip. It may also be more vacuolated. In the apex of the moss *Polytrichum*, and the fern *Pteris*, measurements from sections have shown that the apical cell is much the largest cell, is the most vacuolated, and also has the largest nucleus and nucleolus (Hallet 1978, Michaux-Ferrière 1981b). These features are consistent with apical cells being usually found in the G_2 phase of the cell cycle, rather than in G_1 as the derivative cells are (see Chapter 4, Table 4.3), but this may not entirely account for the large size and vacuolation of the apical cell. The typical apical cell does not, however, owe its size to being polytene or polyploid – it is diploid, at least as long as it is functioning as an initial (see Section 1.1.2). As shown in the fern *Pteris*, the apical cell as well as being meristematic is also differentiated compared with the neighbouring cells, having a large vacuole, weakly staining cytoplasm, and numerous starch grains and lipid granules (although this may be only in the older apical cells) (Michaux 1968). The apical cell and its neighbours are all typically meristematic, in having a large nucleus and nucleolus and relatively simple mitochondria.

In *Polytrichum* and *Pteris* the apical cell has about the same concentration of cytoplasmic RNA as its neighbours but because of its large size it is the cell with most RNA (Hallet 1978, Michaux-Ferrière 1981b). The synthesis of RNA in *Polytrichum* was followed by pulse-chase labelling with [^{3}H]uridine. The change in distribution of the label with time showed that the label was incorporated first into the nucleolus and then passed into the cytoplasm where it accumulated. The relative activity initially incorporated into the nucleoli of the different cells of the apex was proportional to nucleolar volume, so the apical cell appeared to synthesize RNA more rapidly than the other cells because of its larger nucleolus (Hallet 1978). In similar experiments with *Pteris*, similar results were obtained (Michaux-Ferrière 1981b). The apical cell resembles the cells of the central zone of some higher plant shoot meristems (see Chapter 5) in showing greater vacuolation than the more peripheral cells derived from it, but having a similar activity in RNA metabolism.

1.2.1.   *What determines the division sequence?*

Cells are cut off from the three basiscopic faces of the apical cell in apparently strict sequence. This can be inferred from the positions of the new cell walls as seen in transverse sections. What positions the cell wall? It might be assumed that a preprophase band of microtubules forms in the apical cell, but direct evidence is lacking (Christianson 1996). Even so, it is not known what controls the positioning of the preprophase band. Nor is it understood what determines the order in which cells are cut off from the apical cell or the subsequent sequence of divisions of the daughter cells. As far as it is known, for a given apical cell the sequence of divisions is just as likely to be clockwise as anticlockwise, but once the handedness is established it seems to persist: the later divisions are therefore almost certainly influenced by the positions of the earlier formed cell walls. This may be especially so in the asymmetrical divisions of the apical cell in those mosses with a helical leaf arrangement.

There seem to be no detailed ultrastructural studies of the apical cell in relation to cell division. Hébant, Hébant-Mauri and Barthonnet (1978) note that in mosses polarization of the cytoplasm was associated with division: a zone of dense cytoplasm formed on the far side of the cell from the new cell division so that each new derivative cell formed by division of the apical cell was formed from the more vacuolated side of the apical cell. The division of the apical cell is thus an example of unequal division, which is typical of formative cell divisions (Gunning, Hughes and Hardham 1978). Presumably the nucleus of the apical cell moves around the cell to the new division face, but detailed data are lacking.

In the fern *Onoclea*, the newest apical cell derivative always seems to be formed (cut off) by a division parallel to the shortest side of the apical cell (Lintilhac and Green 1976). Since the segmentation pattern implies that successive divisions are from successive faces of the apical cell, this suggests that apical surface growth is non-uniform. The side of the apex where the most recent apical cell division has occurred, and also to a lesser extent the side where the penultimate division occurred, must grow faster than the rest of the apex. This then suggests a rotation of growth rate around the apex, being slowest adjacent to the face of the apical cell where division is about to take place and fastest on the opposite side of the apex, adjacent to the junction of the two non-dividing sides of the apical cell. Is this perhaps comparable with circumnutation in the higher plant? In the *Onoclea* apex, three domains of wall cellulose reinforcement correspond to the three sets of merophytes formed by the apical cell. In each domain the microfibrils showed horizontal hoop-type reinforcement parallel to the face of the apical cell subtending that sector of the shoot (Lintilhac and Green 1976), consistent with their orientation being determined by lateral tensions produced in the newly formed wall by the faster transverse growth of the newly formed merophyte. However, the *Onoclea* apical cell may not be typical. In the *Equisetum* shoot apex the sides of the apical cell are apparently equal in length (Gifford and Kurth 1983), which implies that the growth rate of all walls round the apical cell is the same, and this may be the more common situation.

In *Equisetum*, the cell cut off from the apical cell then divides by another anticlinal division parallel to the first, and then each of these superimposed cells divides anticlinally but radially (the lower cell first) so that in transverse section a sextant of cells is seen below the apical cell

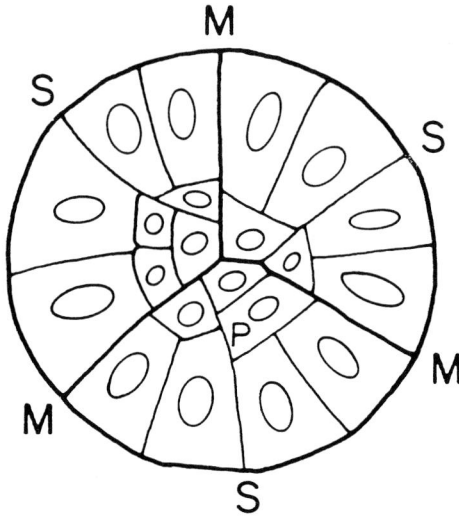

Figure 1.4 Anticlinal divisions in the merophytes below the apex (T.S.) of *Equisetum*. M, boundaries of merophyte sectors; S, sextant walls. From Gifford and Kurth (1983).

(Fig. 1.4) (Gifford and Kurth 1983). The details of the subsequent cell divisions and cell lineages are not known since a precise analysis of the sort made for the root of *Azolla* (Gunning, Hughes and Hardham 1978) has not yet been made for the shoot. The most complete information so far is for cell divisions leading to leaf formation in the mosses (Berthier 1972). Where does the epidermis first become distinct in mosses, just below the apex?

1.3.   The apical cell and organogenesis

The apical cell may not be simply a source of cells. Its structure and functioning may determine the frequency and positions of the organs ultimately derived from its divisions. In the mosses, leaf initiation is very regular and depends on, and can be predicted from, the cell lineage. In those mosses having a wedge-shaped apical cell (shaped like a segment of an orange) cells are cut off from the two basal faces to produce a stem in two sectors, each of which forms leaves (Smith 1955). The leaves are therefore in two ranks, each having been derived from cells formed only from one face of the apical cell. Where the apical cell is tetrahedral the leaves may be in three ranks, as in *Fontinalis* (see p. 422 in Gifford 1983), each sector of the stem giving rise to a file of leaves. Where the leaf

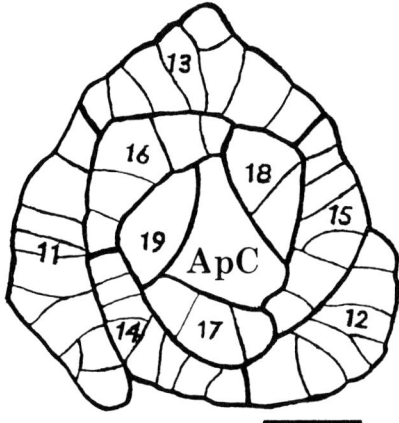

Figure 1.5 Spiral arrangement of the leaves in the moss *Polytrichum* (T.S.). The wall cutting off a derivative from the apical cell is angled to the face of the apical cell so that the merophytes are formed in a spiral, which is then reflected in the leaf arrangement. Bar = 20μm. From Wigglesworth (1956).

arrangement is helical, as in *Polytrichum*, this can be traced to the asymmetrical division of the apical cell. The new cell wall instead of being parallel to the old wall is either at about 20° to the apical cell wall (Hallet 1969) or joins with it at the 'trailing' end (Wigglesworth 1956). In either case this results in a 'rotation' of the apical cell of about 20° for every three divisions (Fig. 1.5). This has also been observed in the moss *Anomodon* where successive divisions of the apical cell are at an angle of > 120° so that an apparent torsion is introduced into the apex, which corresponds to the 3/8 phyllotaxis (Bonnot 1967). In the mosses, therefore, there is a direct relationship between cell lineage and leaf initiation and positioning, with each face of the apical cell forming derivatives which give rise to one leaf per merophyte.

This direct relationship, however, does not seem to hold in more complex plants with apical cells, such as the lycopods, ferns and horsetails. In the fern *Dicksonia* the apical segmentation was observed from reconstructions of the cell pattern of the shoot apex surface. There was no relation between the positions of leaf initiation and segmentation. Furthermore, the helix of segmentation and the helix of leaf initiation were not always homodromous (in the same direction); they were often antidromous (in opposite directions), which is not consistent with there being a correspondence between merophytes and sites of leaf initiation (Hébant-Mauri 1975). Similarly in *Equisetum*, where the three segments of the stem axis can be clearly demonstrated, there was no correlation between

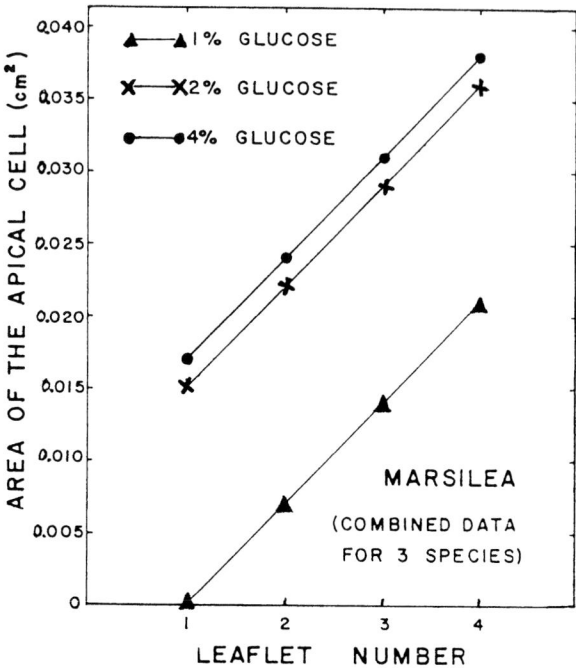

Figure 1.6 Leaflet number in the fern *Marsilea* is a function of the size of the apical cell. The larger the apical cell, the more leaflets are formed. From White (1968).

segmentation of the apical cell and leaf position (Golub and Wetmore 1948). This was because (1) most leaves did not lie within a given segment but were formed across the boundaries of the merophytes, (2) the arrangement of the whorls of leaves at successive nodes was alternate, and (3) the number of leaves per whorl increased and then decreased distally up the stem. A decrease in the number of leaves per whorl in *Equisetum* was correlated with a decrease in stem diameter up the stem and a corresponding decrease in apex diameter during development (Bierhorst 1959). In *Selaginella* there was also no direct correspondence between the pattern of merophyte formation and the pattern of leaf formation (Jernstedt, Cutter and Lu 1994).

However, there is a linear correlation between apical cell area and leaflet number in *Marsilea* (White 1968). It is not known whether this is a function of the number of cells involved in making the primordium, or of the mean cell size of a fixed number of cells, in the sector adjacent to one face of the apical cell. This is not a relationship depending on the absolute apical cell area, but the relative area, since the absolute area depends on the culture conditions used, the apical cell being larger for any given leaflet number when glucose is present (Fig. 1.6).

There seems to have been an evolutionary jump from mosses to ferns, from a cell lineage mechanism for determining leaf formation to a multi-cellular or tissue-level mechanism in the ferns and higher plants. This might lead us to search for a coincident and equivalent evolutionary jump in some aspect of physiology, for instance the formation of some particular type of growth substance or a new mechanism for growth substance action.

1.4.   Origin of the apical cell in early development

Not all of the cells of the plant body have been formed from the apical cell, however, because the apical cell itself arises in early development from one of the cells at the tip of the young apex. The origin of apical cells has been studied in callus derived from a portion of the adult shoot meristem of *Pteris cretica* (Michaux-Ferrière 1975) although not in sufficient detail to see exactly what happens at the cytological level. The callus forms nodules in which a zone of peripheral cells becomes distinguished by the large nuclei and an increased mitotic index (14.8%). Continuous [^{3}H]thymidine labelling suggests a cell cycle of about 20h and cytophotometry shows that the cells are mostly in the G_1 phase. Then, as the apical cells are becoming distinct, the G_2 phase of the cell cycle predominates. In the neighbouring cells of the callus G_1 predominates. This implies either some degree of synchronization of the cell cycles of the developing meristematic cells or else G_2 becomes the longest part of the cell cycle in the apical cells, but G_1 does so in the subjacent cells. One of the more central cells of the newly formed apical meristem then becomes differentiated from the others, develops the characteristic cytology and becomes the apical cell. This suggests that the shape and position of the apex determine which cell becomes the apical cell and that the apical cell therefore arises because of its position – it happens to be the right cell in the right place at the right time.

2

A Source of Cells: the Meristem

2.1. The apical meristem as a source of cells

In higher plants (gymnosperms and angiosperms) and in some pterido-phytes such as *Lycopodium*, *Isoetes* and *Selaginella* (Eames and MacDaniels 1951), a single apical initial cell cannot be identified. Instead, the initials are part of a multicellular meristem at the tip of the shoot (Fig. 2.1). The apical meristem cells seem indistinguishable from each other. But are all the cells of the meristem equivalent and equally able to act as initials, or are some cells more important than others and do some act as permanent stem cells (Francis 1997)? Are there, in fact, a few cells that are the permanent initials, and so the progenitors of all the rest, but are not clearly differentiated from the other cells as the apical cell is? In addition to providing new cellular material, does the shoot apical meristem also have a role in organogenesis?

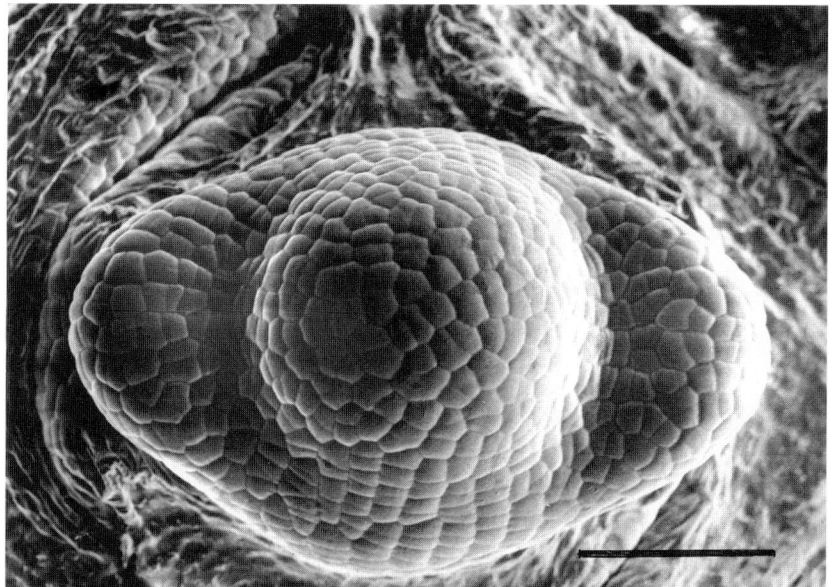

Figure 2.1 Apical dome of *Silene coeli-rosa* seen from above. The youngest pair of leaf primordia are on opposite sides of the apex. The summit of the apex is a mosaic of cells with no single cell different from the others. Bar = 50μm.

2.2. How many initial cells and how long do they persist?

In theory there could be still a single cell, but indistinguishable from the rest, that always remains at the tip of the meristem and gives rise to all the rest, as a true apical cell does. Alternatively there could be a group of two or more cells whose walls meet at the tip of the shoot apex. The point at which the cell walls meet would have to remain at the exact tip or else it would become displaced down the apex and some of the cell group would be lost as initials. As long as the junction of these cells were to remain less than one cell's breadth from the tip then the cells could maintain their apical position and act as initials. Presumably there are several cells at the summit of the shoot from which all the rest are derived.

2.2.1. Evidence from surface displacement

Growth at the shoot tip has been examined by continued direct microscopic observation of the changing cellular configurations on living apices

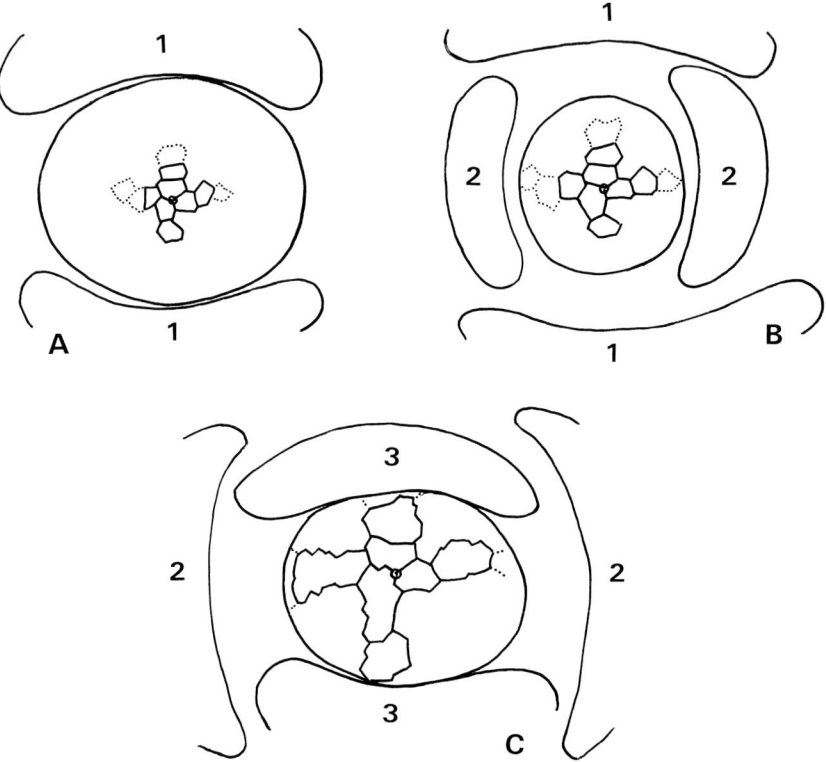

Figure 2.2 Displacement of cells at the summit of the shoot apex of *Anagallis* as traced from successive images of the same apex at intervals during its growth. This is achieved by making a dental impression-polymer mould of the apex, which is removed, allowing the apex to continue growth. The mould is then used to make an epoxy resin replica of the apex, which is then photographed in the scanning electron microscope (SEM). Another mould is then made later and the process repeated. The outlines of cells on the photographs of the replicas can then be traced. The diagrams show the changes in size and position of a group of cells (and the clones derived from them by normal cell division) through three plastochrons (A, B, C) and the production of three pairs of leaves (1, 2, 3). The three abutting cell walls at the summit of the apex (circle) are still there three plastochrons later. From Green, Havelange and Bernier (1991).

(Newman 1956, Ball 1960a) or by placing marks on the apical surface and observing their displacement (Wardlaw 1949, Loiseau 1962, Soma and Ball 1963, Bergann 1965, McAlpin and White 1974). More recently, Green, Havelange and Bernier (1991) have used the extremely effective and elegant technique of making successive replicas of the surface cell pattern of the same apex as it grew (Fig. 2.2). All these observations showed that while other cells were displaced down the apical dome, there were apparently three cells at the tip of the meristem which persisted there during the period of observation and which could therefore be regarded as the initials. However, there are bound to be always one or more cells at the summit of the apex. The real question is do these apparent initials always remain at the tip of the meristem or can they be displaced by adjacent cells which then become initials? The cell packets seen on the surface of the *Phaseolus* seedling apex appeared to be similar and suggested that all cells were dividing and potential initials (Gavaudan and Gastelier 1970). Are all surface meristematic cells equally qualified to be initials and can they become initials eventually?

When the initial cells of the *Lupinus* shoot meristem were inactivated by pricking the summit of the apex, the wound was gradually displaced off the apical dome, implying that other cells took over since the dome continued to function (Soma and Ball 1963). However, it could be argued that it is difficult to prick the exact centre of the dome and that all the punctures were inadvertently made just to the side of the functional centre, which may not have coincided with the geometrical centre.

The direct observations necessarily lasted only two or so plastochrons. For how many plastochrons can initial cells remain at the apex? Spots of carbon black, placed over the surface of the apical dome of *Lupinus*, were mostly displaced away from the summit by growth of the apex (Soma and Ball 1963). Some marks were not displaced, but these became fewer with time, until by 42 days only 6.7% had still not moved. Presumably these were the marks that had been placed on the actual initials, which had therefore persisted for at least 13 plastochrons.

The persistence of cells as apical initials can be followed over longer periods by the use of chimeras and by the technique of clonal analysis where it is possible to mark individual cells and follow their subsequent long-term fate.

2.2.2.   *Clonal analysis*

A chimera is a plant consisting of two or more clones of cells differing in their genotype and phenotypic expression. When different layers consist of different clones (e.g. epidermis and outer cell layers of one clone, inner cells of another) this is described as a mericlinal chimera. Where radial sectors of the plant axis are different clones, the chimera is said to be sectorial. Mericlinal chimeras may show sectored expression in which, for instance, a stripe of cells down the plant lacks chlorophyll or other pigmentation, such as anthocyanin. The minimum number of sectors observed corresponds to the maximum number of initial cells at the shoot apex. Often mericlinal chimeras occupy one-third of the stem, implying three initial cells (Neilson-Jones 1969; Tilney-Bassett 1986), thus corroborating the direct observations on apices. No mericlinal chimeras have been recorded for ferns (Bierhorst 1977), this being consistent with there being only a single initial cell (see Chapter 1).

Chimeras become more useful when they can be induced at known times during development of the meristem so that the fate of their derivative cells can be followed during subsequent development of the plant. This can be done experimentally by inducing mutations in apical meristem cells so that artifical chimeras are produced. The study of the fate of the mutagenized cells, and the clones derived from them, is clonal analysis. In plants this is possible because during development and displacement the cells do not move position with respect to their neighbours. In this way, information can be obtained about the number of initials in the shoot apex, how this number changes during development, how long they operate as initials and what organs and tissues they contribute to, so giving information about the structure of the apex as a source of cells.

When mutations are induced experimentally in the initials or their derivatives at known stages in the development of the meristem, clonal analysis can tell us about meristem structure and functioning. The positions of the cells in the meristem at the time of their transformation can be inferred from the number and size of the clonal sectors: the fewer and larger, the earlier the clonal sectors were formed. Knowing the cellular strucure of the apex it can be deduced which cells in the apex these were at the time of treatment. The most useful technique is to use heterozygous plants for a dominant trait such as anthocyanin production. When such plants are then exposed to a mutagenic treatment, it requires the knocking out of only the one copy of the anthocyanin gene in the genome to produce a colourless cell and a colourless clone from its derivatives. The number, position and fate of the transformed cells depends on when the treatment to give clonal differences is given. Thus the time at which cells have their fate sealed or restricted (but not necessarily determined) can be found by doing the mutagenizing experiment at different times during development. Early in embryogenesis a small number of apical initial cells may contribute to many tissues and organs. As cells are displaced by growth down the apex they are restricted by their position so that they have the potential to contribute to fewer organs or smaller parts of them (Poethig 1987).

Chimeras for chloroplast deficiency were induced experimentally by a mutagen (ethyl methane sulphonate) in the shoot apices of the seeds of several species, including monocotyledons and dicotyledons (Dulieu 1968, 1969). In all cases the resulting chimeras gave leaves with clones of basically longitudinal sectors. These were consistent with the leaf originating from a group of cells on the apical surface, each of which gave rise to a clone contributing to part of the final leaf structure. The limited persistence of sectors vertically down the plant suggested that there were no permanent initials in the apical meristem but that all cells had the potential to be displaced down and into the leaves. The tendency for this to happen was inversely related to the length of the mutant sectors, longer sectors being rarer than shorter ones. A sector that persisted for only a few nodes indicated that the initials that formed it were quickly displaced from the meristem, whereas a long sector indicated relatively long persistence of the initial cells in the meristem. Dulieu (1969) pointed out that the persistence of cells as initials would depend not only on the rates of division in the apical meristem (where it is usually relatively slow at the apical summit – see Chapter 3), but also on how much of the meristem was used up in formation of the primordia and hence how quickly, in terms of plastochrons, the apical dome was reconstituted – in other words, how the apex is partitioned (Chapter 8). In a mature tobacco apex making relatively small primordia, the rate of replacement of the apical dome was slower, and hence the initials persisted longer than in an apex such as *Vicia* in which the primordia are larger relative to the apical dome (Dulieu 1969). Dulieu also pointed out that in a young apex in which growth and division is more prevalent than at the summit of a more mature apex, the displacement frequency will be higher. The persistence of initials will therefore tend to be greater in more mature apices until flowering is approached when division and growth rates in the apex increase once more (Chapters 4 and 9).

Initial cells can remain as initials for very long periods in some cases. In *Zea* (maize), initial cells at the apex can contribute to both the vegetative parts of the plant and also to the reproductive parts, the ears and tassel (McDaniel and Poethig 1988). In *Helianthus* (sunflower), too, the same initials can give rise to many internodes and also to flowers (Jegla and Sussex 1989). This is despite the finding that on a shorter time scale the central zone of the *Helianthus* apex appears not to act as a set of initial cells (Steeves et al. 1969). This shows that the time scale of the experiments matters for their interpretation and, of course, ultimately all cells are traceable back to the zygote.

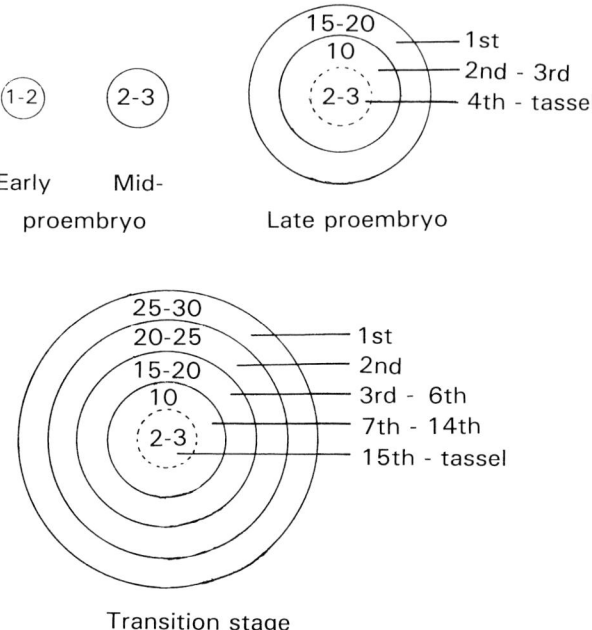

Figure 2.3 As the shoot apex of maize (*Zea*) grows from the proembryo to the transition stage in the seed, increasing numbers of layers of cells develop. The diagrams show the apex as though viewed from above. The numbers in the rings indicate the number of cells in each layer at each developmental stage. The cell layers giving rise to the 1st–15th internodes and the tassel (the male inflorescence at the tip of the mature plant) can be identified. The two to three cells remaining at the summit of the meristem give rise to only the last internode and the tassel. The whole of the shoot is derived from the two to three cells comprising the apex at the early and mid-proembryo stages. From Poethig, Coe and Johri (1986).

Poethig, Coe and Johri (1986) were able to construct fate maps for the cells in the apical meristem of *Zea*. Before initiation of the shoot apical meristem there appeared to be two to three cells in the central-most ring at the summit of the embryonic apex, and these are presumably the initials that can contribute to both the vegetative and reproductive shoot. The transition stage of embryo development is when the major organs, and the root and shoot meristems are beginning to become apparent. At this stage clonal analysis was consistent with the epidermis of the shoot meristem consisting of 72–88 cells arranged in five concentric rings (Fig. 2.3). This corresponds to what is seen by microscopic observation of the embryo: the epidermis of the apical dome (the region distal to the coleoptile) does indeed consist of about five rings of cells, and the total number of all cells in the apical meristem at this stage would be about 150–200. These seem to be the cells that act as initials for the earlier-formed organs. The further down the apical dome, the shorter were the sectors of the plant to which the initials contributed (Poethig, Coe and Johri 1986). After the transition stage in the embryo all that the cells at the summit of the apex seem to be doing is acting as a source of cells that contribute to the tassel and which displace the other progenitor cells down and into the organ-forming regions on the flanks of the meristem.

It is worth noting that clonal analysis tells us about the sources of cells, the number of initials and cell fate, but it does not tell us about when the cell fate becomes sealed, i.e. it does not indicate when the cells become determined. Presumably the more the cells are displaced away from the apical summit, the more restricted is the fate of the initial cells. However, it must not be assumed that the cells derived from the meristem were necessarily determined as anything more precise than shoot cells, although it is known that the fates of the cells become fixed (because it can be predicted where they will end up in the tissues of the mature plant).

At the transition stage of maize embryo development, most of the cells in the shoot meristem had probably become determined as non-initial cells because twinned shoots were produced only by irradiation of embryos before the transition stage (Poethig, Coe and Johri 1986). In the absence of experiments to change their fate, it can only be said that the cells are specified, i.e. their fate as particular cell types is prescribed, but not that they are determined, i.e. unable to follow any other fate if circumstances were to allow it. This emphasizes the hierarchical nature of determination. Although the cells of the shoot meristem may soon become determined as cells other than initials, they may not necessarily be determined as cells of particular organ or cell types even though that is where they end up.

So far, clonal analysis has revealed the essential structure of the embryonic apical meristem so that the positions of the cells within it that will give rise to the various parts of the adult plant can be deduced. How invariable is this? Maize and sunflower are determinate plants in that they form a relatively fixed number of leaves before the terminal inflorescence is formed, at least in the conditions under which they have been grown for these clonal analysis experiments. Are there meristems in which the fate of the cells is not fixed early on, as these appear to be? In *Arabidopsis*, a less obviously determinate plant, most of the cells of the shoot meristem give rise to the first six leaves whereas the rest of the shoot, whether it forms only one or two more leaves or 40 leaves before flowering, is all derived from only a few cells at the summit of the meristem (Irish 1991).

For meristems, as for shoots with an apical cell, it can be asked: are the initials just a source of cells or are there particular initials and cell lineages for particular cells or tissues or organs? Clonal analysis shows that the fate of cells does not depend on lineage because different parts of the plant and different organs can arise, to varying extents, from the same initials in the embryonic apex. Similarly, some clonal sectors can extend through vegetative nodes and leaves and into the inflorescence and flowers, showing that the same initials can contribute to the vegetative and reproductive parts of the plant, and so there is no germ line and no 'méristème d'attente' or 'waiting meristem' at the summit of the apex consisting of initials used only for reproductive structures (see Chapter 3, Section 3.2.1). Cells that are initials early in development can contribute to a lot of the plant's structure, but those that are initials later contribute less.

Clonal analysis, however, can tell us only about the fates of cells at the time of mutagenic treatment, which is usually at various stages of embryo development in the seed. To understand the functioning and structure of the more mature meristem, treatment for subsequent clonal analysis needs to be done on the meristem at the stage selected for investigation. With the exception of the work of Dulieu (1969), clonal analysis does not seem to have been done with the specific aim of understanding the changing structure of the shoot apical meristem. The information available has been derived from analysis concerned mainly with leaf initiation and formation (Poethig 1984, 1987).

How many initials are there? The answer (for the surface) is very few, probably three, determined by the difficulty of maintaining a single point always exactly at the apex, but these cells can probably be displaced during the development of the plant. Although in some plants, such as sunflower, there are initials that can give rise to much of the plant, there seem to be no initials that give rise to the whole plant, and so there are no permanent initials. What is known is consistent with those cells that are initials being so by virtue of their position, not their cell lineage. The functioning of the apical meristem in the vascular plants seems to be a result of positional information in the meristem as a whole, with the cellular structure being a consequence of this functioning rather than its cause. The nature of the positional information is still in question.

2.3. Numbers of initial layers

So far, only the initials on the surface of the meristem have been considered. Most meristems are more complicated than this. Meristems may be classified into basically three types: monoplex, simplex and duplex (Cutter 1971). In monoplex and simplex apices there is a single, superficial, initial layer. In a monoplex apex there is a single initial cell, so that a single division gives both surface extension and increasing bulk, as happens when the apical cell divides. In a simplex apex the superficial cells may divide either anticlinally (normal to the surface), giving increased surface, or periclinally (parallel to the surface) to produce cells internally. The simplex apex is characteristic of the gymnosperms. In the duplex apex, characteristic of the angiosperms, there are at least two, sometimes more, initial layers. The outer, superficial layer or layers divides only anticlinally to give an increase in the surface layers, whereas all the inner cells are produced from subepidermal cells in which divisions are periclinal as well as anticlinal. The outermost layer, the epidermis, therefore remains a separate tissue with its own initials and does not normally contribute to the underlying layers of the shoot apex.

2.4. Meristems with only one set of initials: mantle and core

In gymnosperms, a longitudinal section of the shoot apex typically shows evidence of periclinal divisions in the surface cells so that the epidermis at the apex and where the primordia are initiated is not clearly differentiated from the subepidermal cells, although down the flanks of the apex and in the rest of the plant it becomes so (Fig. 2.4). This simplex structure is confirmed by the unstable nature of chimeras in the gymnosperms, specifically in some conifers. They were originally thought to be unstable sectorial chimeras but have been shown to be mericlinal chimeras. They tend either to take over the whole of the shoot or to disappear, reverting to the non-chimerical structure typical of the outermost layer of cells (Tilney-Bassett 1986). This is consistent with the histological evidence of occasional periclinal divisions in the outermost, L1, layer in the conifer apex which contribute cells to the inner, L2, layer. A mutation in the epidermis to form a mericlinal chimera can therefore be propagated into the underlying cells and eventually to all the apical meristem cells. There

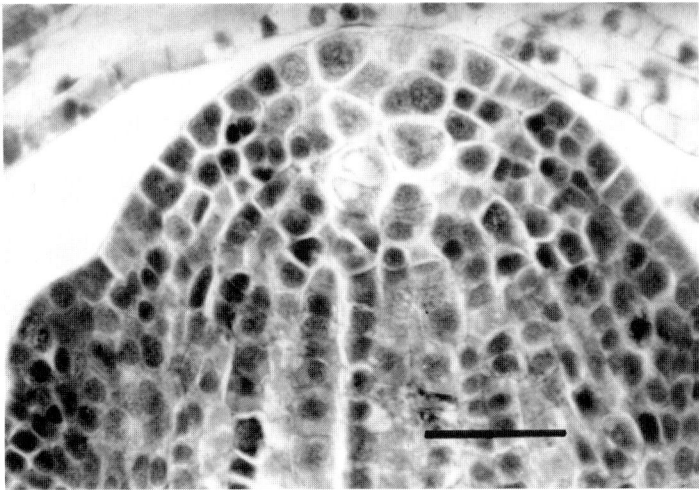

Figure 2.4 Shoot apex of pine (*Pinus*) (L.S.) showing the lack of tunica–corpus structure, and the presence of large central mother cells which have thick walls and are presumed to be dividing more slowly than the cells around them. Differentiation of vascular tissue is beginning at the lower end of the pith-rib meristem, below the central mother cells. Bar = 50μm.

are only two layers, L1 and L2, but as shown by the lack of persistence of the chimerical structure they are only temporarily independent. However, this does indicate the beginnings of a layered meristem, and so is intermediate between the ferns and the angiosperms.

In *Cupressus* (a gymnosperm) the periclinal divisions in the apical cells at the summit of the meristem are most frequent during the summer and are much less frequent or absent during the winter (Pillai 1963). It is not clear whether this is just a function of a lower frequency of divisions in the winter. Even if anticlinal divisions were not happening this would not be obvious since the cell pattern would not change, whereas periclinal divisions just would not be apparent if there was no or little division. The periclinal divisions in the outer layer are confined to the apical summit (only one to three cells as seen in longitudinal section) and do not occur on the flanks; where the leaf primordia form, the epidermis is distinct.

This simplex structure of the gymnosperm shoot apex is sometimes described as mantle and core. Often at the apical summit there is a group of larger, underlying cells that may have thick walls, suggesting slow growth and infrequent division (Fig. 2.4). These central mother cells therefore look as though they are initials, but there are no quantitative data on division and growth rates at the summit of such apices to allow further interpretation. Also there is no quantitative information on the relative frequencies of anticlinal and periclinal divisions in simplex apices so it is not known whether divisions are in all planes, as they apparently are in the corpus in angiosperms.

The orientations of microtubule arrays in the cell cortex tend to be arranged perpendicularly to the axis of cell expansion (Green 1980, Gunning and Hardham 1982). In gymnosperm shoot apices the microtubule orientations correlate with the mantle–core structure. The microtubules are arranged randomly as well as anticlinally in the surface cells of gymnosperm apices, randomly in the underlying cells and transversely in the rib meristem (Sakaguchi, Hogetsu and Hara 1990).

2.5. Meristems with a layered structure: tunica–corpus

In angiosperms the shoot apex has a duplex structure and is therefore layered. An outer layer, the tunica, which is one or more cells thick, can be distinguished in which the cell divisions are exclusively anticlinal, i.e. perpendicular to the nearest surface. If the tunica is only one layer thick then it is synonymous with the epidermis. The tunica encloses all the inner cells, or corpus, in which divisions are in all planes, including periclinal (parallel to the nearest surface). The tunica layers grow only in surface area whereas the corpus cells grow in all directions to increase the bulk of the inner tissues. The angiosperms therefore differ from gymnosperms in having an outer tunica of cells in which divisions in the plane parallel to the surface are essentially absent. Divisions in all planes in the corpus can be regarded as being the norm, whereas the tunica is under some form of constraint which prevents periclinal divisions. Unlike the gymnosperms, in angiosperms the orientation of microtubules is exclusively anticlinal in the tunica layers, at least at the summit of the apex (Fig. 2.5). The angiosperm apices are otherwise similar to gymnosperm apices in having random microtubules in the corpus and transverse microtubules in the rib meristem (Sakaguchi, Hogetsu and Hara 1990). The microtubule orientations are again perpendicular to the expected axes of growth of the various cell layers and correlate with the tunica–corpus structure.

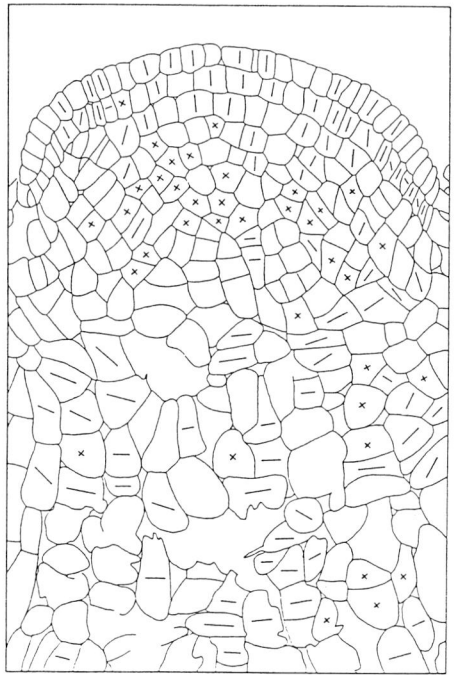

Figure 2.5 Orientation of microtubules in cells of the shoot apex (L.S.) of *Vinca*. The microtubules are predominantly anticlinal in the tunica, except on the flanks where a primordium is about to form, where they are also oblique or periclinal. They are random in the corpus (approximately the 4th–8th layers below the apical summit), and transverse in the rib meristem, below the corpus. From Sakaguchi, Hogetsu and Hara (1988).

2.5.1. Evidence for tunica–corpus structure

The criteria used to recognize the existence of tunica and corpus are the planes of cell division in the apical dome as seen in longitudinal sections. Those surface cell layers in which only anticlinal divisions are observed are classed as tunica. Only the structure of the actual summit of the apex is considered, not the flanks, because on the flanks the tunica structure disappears by the occurrence of periclinal divisions associated with the outgrowth of a primordium.

Chimeras have been useful in providing information about the numbers of initials and layers in the shoot apex. The most interesting chimeras in establishing tunica–corpus structure as a reality have been cytochimeras. In *Datura* these were made by treating seeds with

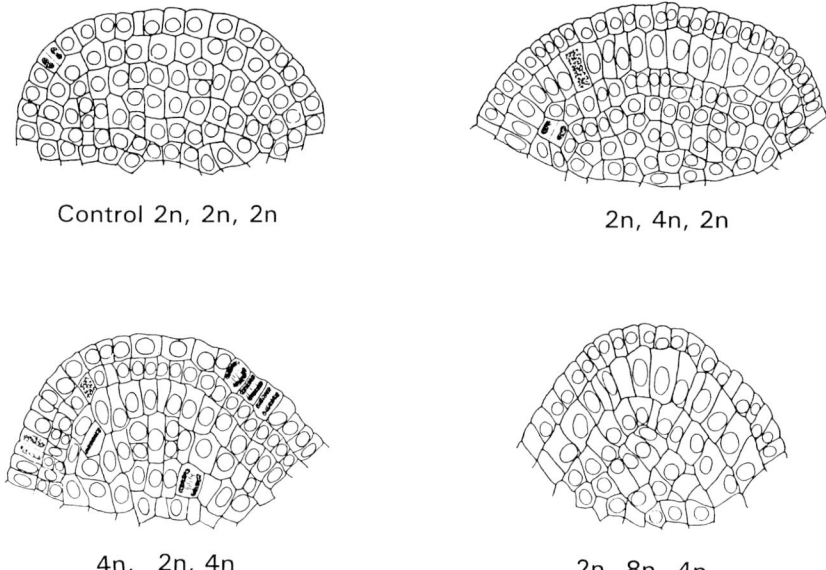

Control 2n, 2n, 2n

2n, 4n, 2n

4n, 2n, 4n

2n, 8n, 4n

Figure 2.6 *Datura* shoot apices (L.S.) showing cytochimerical structure induced by colchicine, and indicating two tunica layers over the corpus. From Satina, Blakeslee and Avery (1940).

colchicine to cause endopolyploidy in some of the cells. The resulting progeny of the endopolyploid cells could be distinguished from each other by virtue of their sizes, tetraploid (4n) and octoploid (8n) cells being visibly larger than diploid (2n) (Satina, Blakeslee and Avery 1940). Sections of the apices of the resulting plants showed various combinations of diploid and polyploid tissues, but all were consistent in showing the *Datura* apex to have three layers, corresponding to two tunica layers and a corpus (Fig. 2.6). These three initial layers are often designated L1 (outermost), L2 and L3.

In higher plants the establishment and persistence of periclinal chimeras (Fig. 2.6) implies (1) that the apex can be layered, and (unlike gymnosperm apices) these layers can remain distinct, and therefore there must be at least as many sets of initial cells as there are layers, and (2) that the initial cells for all the layers must be at the summit of the meristem, otherwise the layered structure would not persist but would be displaced downwards and away from the apical summit. Periclinal chimeras, including cytochimeras, confirm the tunica–corpus structure. What is not certain is whether chimeras with their adjacent layers of

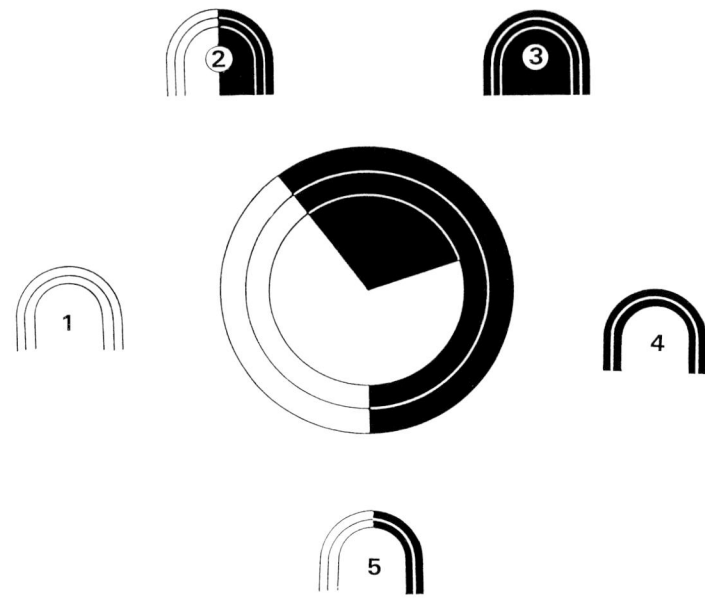

Figure 2.7 A chimerical apex (seen from above as a circle) consisting of two cell clones (white and black) would give rise to buds of uniform, but different clones (1 and 3) or to chimeras which are sectorial (2), periclinal (4) or mericlinal (5). From Kirk and Tilney-Bassett (1978).

different genetic constitutions are more stable than normal, non-chimerical plants in which all layers have the same genetical constitution.

The existence of sectorial and mericlinal chimeras (Fig. 2.7) means that there are at least as many (but there may be more) initial cells, or sets of initials, in each layer as there are sectors in the chimera. Such chimeras have indicated the possibility of anything from one set of apical initials, when the whole apex is chimerical and does not show sectoring (although this could be because there are several sets of initials, all of which have mutated), to as many as twelve sets of initials (Tilney-Bassett 1986). Mericlinal chimeras occupying one-third of the plant indicate three, or possibly six, sets of initials. Mericlinal chimeras occupying one-half (or one-quarter) of the plant indicate the probability of only two (or four) cells happening to sit at the summit and acting as initials. However, unless the chimerical sector can be traced back to a single cell at the apex, it will always be uncertain how many initials there are at the summit of the apex. What chimeras do show is that in the higher plant apex there is not just one but several initial cells and initial layers.

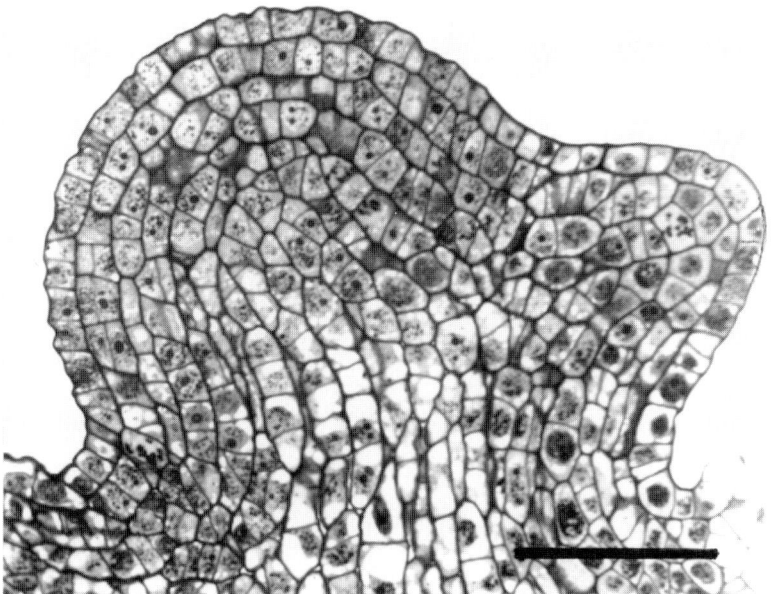

Figure 2.8 In the pea (*Pisum*) shoot apex (L.S.) the three-layered tunica becomes disrupted by periclinal divisions where a primordium is about to form (left). In an existing primordium (right) the increasing contribution of the corpus is evident. Bar = 50μm.

2.5.2. Origin of organs from L1, L2 and L3

Most dicotyledonous apices have a tunica–corpus arrangement with a tunica of two to four layers. Where there are two tunica layers and a corpus these would correspond to the L1, L2 and L3 layers described from chimeras. Although the leaf primordia are superficial in origin, they are formed from all tunica layers and some corpus (Fig. 2.8). However the contribution of L1, L2 and L3 to the mature leaf structure is far from constant even in the same plant, as shown by the varying patterns in variegated plants in which the variegation pattern does not necessarily correspond to the layers from which the tissues originated (Tilney-Bassett 1986). The cell layers from which leaves originate can most easily be seen in those periclinal and mericlinal chimeras in which one or more initial layers is mutant for chlorophyll production (Tilney-Bassett 1986). Because the epidermal cells (apart from the guard cells) do not contain chlorophyll they and any cells derived from them appear colourless. In a white–green–white (W-G-W) chimera the outer tissues (derived from the

Figure 2.9 Contribution of the various layers, L1, L2 and L3, to the leaf varies from leaf to leaf in successive leaves of the same plant of the green–white green chimera, *Ficus benjamina* var. *variegata*. The outer layer, L1, appears white because it is epidermal in origin and therefore only the stomatal cells possess chloroplasts.

epidermis, L1) appear white, the tissues derived from L2 are green and those from L3 are white. It is typical of leaves of such plants that the relative contribution to the mature leaf by the different initial layers varies from leaf to leaf (Fig. 2.9). The form and structure of the leaf is, however, constant between leaves, showing that the developmental fate of the cells depends on their position in the developing leaf and not on their cell lineage. Leaves typically originate from all three layers or sometimes only from L1 and L2, while internal tissues such as the sporocytes of the anther and the ovule tend to originate from L2 (Tilney-Bassett 1986). Thus, periclinal and mericlinal chimeras cannot be propagated sexually because only the genes of the L2 layer will be represented in the gametes. The layers of the apex are therefore best viewed as sources of cells, the fates of which become determined as a function of positional controls during their development and not as a result of their cell lineage. The tunica–corpus structure does not, therefore, seem to have any significance as far as the cell lineages in organ formation are concerned. What, then, is the significance of the tunica–corpus structure?

In the tomato mutant, *fasciated*, the meristem is larger than in wild-type plant and the number of carpels in the flower is increased from four to 17. In chimeras in which the L3 layer of the meristem was *fasciated* but L1 and L2 were wild-type, the meristem was also larger than the wild-type and so was the number of carpels, which numbered 12 (Szymkowiak and Sussex 1992). Similarly, when chimeras were formed from a line of *Lycopersicon esculentum* with five carpels, and *L. peruvianum* with two carpels and a smaller meristem, and when only the L3 layer in the chimeras was from *L. esculentum*, the carpel number in the chimeras was four, and the meristems were the same size as in *esculentum*. In these chimeras the size of the meristem and the carpel number was therefore determined by the L3 layer. In *Camellia*, the formation of stamens and carpels depends on the epidermis. *Camellia* + 'Daisy Eagleson' is a graft chimera of *C. sasanqua*, which forms normal flowers, and *C. japonica*, in which stamens and carpels are missing. The chimera, in which the epidermis (L1) is from *sasanqua* and the inner layers are from *japonica*, forms stamens and carpels, so that the formation of reproductive structures depends on the nature of the epidermis (Stewart, Meyer and Dermen 1972).

These examples show that the number and nature of the organs formed can depend on different layers of the meristem but, since all layers are involved in the construction of the organs, there must be inductive signals passing from layer to layer. And of course in gymnosperms in which there are no layers, organ formation is similar to that in angiosperms with layered meristems. The expression of traits specific to one parent in a chimera may simply indicate that wherever developmental signals originate in the meristem, all cells can react as a function of their position. While the existence of tunica–corpus structure makes possible the construction of chimeras to demonstrate the roles of different layers, this structure does not seem necessary in order for these roles to be fulfilled.

2.5.3.   *Variability of tunica–corpus structure*

The number of tunica layers can be defined as the number of surface cell layers at the apical summit in which periclinal divisions do not occur. The problem is that if periclinal divisions do occur, but infrequently, then the more apices, and the more sections for each apex, that are examined, the fewer the number of tunica layers that will be found. Evidence for a stated number of tunica layers must always be with the tacit proviso that this number is valid only for that specific set of observations. It can only be emphasized what Clowes (1961) pointed out, that the more assiduous the search for periclinal divisions, the fewer tunica layers can be defined in any given species.

The number of specimens that can be examined by sectioning is relatively limited. Very infrequent periclinal divisions would probably not be detected. However, analysis of mericlinal chimeras in a variety of species over 20 years showed that periclinal divisions did in fact occur, although rarely, in the epidermal initials. In Californian privet (*Ligustrum ovalifolium*), there were 22 periclinal divisions (as shown by the occurrence of 22 mericlinal chimeras) in about 5000 divisions of the meristem initials, a frequency of 0.4% (Stewart and Dermen 1970). It was calculated that to observe this number in sections would have required approximately 400,000 medial longitudinal sections. Rarity of periclinal divisions does not equate with absence. Even these rare divisions clearly altered the long-term structure of the meristem. It therefore makes sense to define the number of tunica layers only in terms of what is usually found, with the proviso that periclinal divisions may occasionally be found in the tunica. Only in permanent mericlinal chimeras can there be a fixed number of tunica layers.

The number of tunica layers in a given species may vary with the stage of the plastochron. Often there are more tunica layers at the stage of apex maximal area, when the apical dome is at its largest just before a new primordium is initiated, than at minimal area, just after primordium initiation (Reeve 1948, Gifford 1954, Sussex 1955, Lyndon 1976). For instance, in the potato apex the number of tunica layers alternates between two and five (all the apical layers!) according to the plastochron stage (Sussex 1955), and in *Acorus* between three and seven (Kaplan 1970).

This variation with plastochron stage is linked to changes in the size of the apical dome, which is of smallest diameter just before leaf initiation, and this is when it tends to show fewest tunica layers. Apices of grasses tend to be relatively small and narrow and characteristically have only one tunica layer, the epidermis (Sharman 1942, Esau 1965). However, young, small apices of *Saccharum* tended even to lack a tunica altogether, with periclinal divisions in the epidermis (Thielke 1965). Nutrient stress can also lead to a smaller apex in *Silene* and the increased occurrence of periclinal divisions in the outer cell layers and so to fewer tunica layers (R.F. Lyndon, unpublished data). Similarly, γ-irradiation can cause periclinal divisions in outer tunica layers (Crockett 1968).

At flowering, the tunica–corpus structure may become modified (Cutter 1971), but little attention has been paid to exactly how. Whether it changes as a function of the generally increased size of the apex, or the changing shape of the apex, is not known.

2.5.4. What limits the plane of division to anticlinal in tunica cells?

The tunica–corpus concept has been accepted as a way of describing the structure of the apex, but usually (except for chimeras) without any further enquiry as to why it exists at all, or what light it might throw on the control of growth in the apical dome. However, Dermen (1969) compared the structure of apices: those with an apical cell, those without a tunica (gymnosperms), and those with a tunica. He concluded that the plane of division was restricted in tunica cells because in general they were narrow, as in the peach, and at mitosis the metaphase plate could only be accommodated vertically, so giving an anticlinal division. However, this hypothesis does not survive wider observations. In a number of published photographs of apices, as in the pea apex (Fig. 2.8), many of the tunica cells are wide enough to suggest that it ought to be possible for them to divide periclinally.

A possible more general explanation is based on the mechanics of the shoot apex and the stresses within it. The zonate cell pattern of the corpus of some apices can theoretically be accounted for purely by mechanical and geometric factors operating within the apices (Niklas and Mauseth 1980). However, even in these models the prior existence of a tunica was assumed, but perhaps development of these ideas could provide a mechanical explanation for the tunica. The anticlinal orientation of the microtubules in tunica cells (Sakaguchi, Hogetsu and Hara 1990) would be consistent with the idea that the microtubules are orientated normally to the direction of stress resulting from the predominant direction of growth in the tunica (parallel to the surface), and that new (anticlinal) cell walls form in the position of least shear, as proposed by Lintilhac (1984). If it is the epidermis that is controlling growth, as in stem segments (Kutschera, Bergfeld and Schopfer 1987), it is difficult to see why the tunica should be several layers of cells, or why the number of tunica layers should change during growth. However, if growth is controlled by changes in the plastic extensibility of the outer walls of the apex, the existence of several tunica layers suggests that such changes may occur not only in the epidermal walls but also in the walls of the rest of the tunica. Changes in the number of tunica layers during growth might then be indicative of a changing distribution or concentration in the shoot apex of signalling factors which affect wall plasticity. In the pea embryo there are periclinal divisions in the tunica until the apex has initiated two or more primordia (Reeve 1948). This would be consistent with the tunica structure being a function of the presence and growth of the primordia formed on the apex.

2.5.5.   The significance of different numbers of initial layers

In the dicotyledons chimeras typically have a three-layered structure, indicating two tunica layers, whereas monocotyledons frequently show only two layers indicating a one-layered tunica. What is the significance of monoplex structure in lower plants, simplex structure in gymnosperms, and duplex structure in angiosperms, with one tunica layer in many monocots and two or more tunica layers in dicots? Whatever the factors controlling the occurrence of periclinal divisions in the outer layers of the meristem, they would seem to be different, but consistently so, in gymnosperms, monocotyledons and dicotyledons. Several other questions may be asked to which there are no answers as yet. For instance, are there any correlations between the type of layered structure of the meristem and any other features of leaf formation and growth? Are the microfibrillar and microtubular orientations which are correlated with the tunica structure (or lack of it in the gymnosperms) the cause or consequence of the apical structure?

2.6.   Hidden structure: internode and node initials

The meristem not only most obviously produces leaves but also makes other structures, notably axillary buds, nodes and internodes. Where and when do the initials for these structures arise? If they arise in the apical meristem is this reflected in the cellular structure of the meristem?

2.6.1.   *Where do internodes come from?*

Cells of young nodes and internodes can be distinguished from each other clearly by their phenolic contents in a wide range of plants, indicating that nodes and internodes originate from different cell layers in the apex. Distinction into nodal and internodal cells is most easily seen in plants with two or more leaves at each node (Zobel 1989a). In *Sambucus* there are coenocytic tannin cells that extend for the whole length of each internode but are not found in the nodes (Zobel 1985). These coenocytes, which therefore arise in a layer of cells that gives rise to an internode, can first be detected in the ninth cell layer from the surface of the apical dome. This is at the level of the formation of the pith-rib meristem and below the level of the youngest leaf axil. Each internodal initial cell layer is separated from the next internodal layer by at least one layer of nodal cells (Zobel 1985). In *Polygonum*, a layer two or four cells deep corresponding to the length of the youngest internodes could be distinguished (Zobel 1989b). In *Silene*, in the second youngest node/internode at the shoot apex, four cell layers could be distinguished belonging to the node and four to the internode (Lyndon 1987a). In older internodes and nodes, the internodal cells were in obvious files whereas the nodal cells were not. In *Silene* shoot apices, sometimes only two phloem sieve tube elements could be seen extending the length of a young node or internode, suggesting that the node and internode each originated from only two layers of cells (Lyndon 1987a).

These observations imply that nodes and internodes are initiated from successive layers of cells in the shoot apex. Since nodes and internodes are formed only when primordia are initiated, we might expect them to be initiated on the apical dome at the same level as the primordia. Are there any indications of regular transverse divisions in the apical dome which could represent the division of node/internode mother cells? Although periclinal divisions can be found in a number of plants in the apex just under the epidermis, these cannot be definitely equated with the formative divisions for the nodes and internodes (Zobel 1989b) and may simply be periclinal divisions contributing to the growth and increased volume of the apex. In maize, clonal analysis showed that most clones of cells started at the base of an internode and ended in a leaf at the top of an internode (McDaniel and Poethig 1988), consistent with the node and internode originating from a common initial. However, there are presumably more periclinal divisions in the apical dome than are necessary to form nodes and internodes.

If a layered structure arises as the leaves arise, then layering in the shoot apical meristem (except for tunica–corpus structure) may develop only as internodes begin to develop. Let it be supposed that the increase in cell number per plastochron in the apical dome involves the formation of two cell layers (node plus internode initials) per leaf (formed when the dome is at near minimal size) which can then be multiplied by the number of further cell cycles per plastochron in the shoot apex. In plants like *Silene*, forming a pair of leaves each plastochron, there would be four new cell layers formed per plastochron (two nodal plus two internodal layers). This would require two cell cycles starting from an original single mother cell layer. The number of layers would then be increased by the number of cell cycles subsequently occurring during the rest of the plastochron. In *Silene* the number of cell layers found in the axis associated with the youngest primordium was eight, i.e. four nodal cells plus four internodal cells (Lyndon 1987a), which would have required one further cell cycle, or three cycles in all. Since the plastochron is 4 days, and the cell cycle 1 day (Miller and Lyndon 1975, 1976), this would allow four cell cycles, whereas only three would have been required to generate the eight cell layers from a mother cell. The one cell cycle per plastochron remaining could have been when the mother cell itself was formed.

These observations need to be extended by identification of genes expressed specifically in internodes so that the cell differentiation and determination within the shoot apex can be followed more precisely at the molecular level.

2.6.2.   *Early growth of stem frusta*

The cells that eventually form the node and internode are contained within the cylinder of axial tissue bounded by the upper and lower limits of the primordial bulge, i.e. between successive leaf axils. Since this axial tissue is usually wider at the base than at the top, because of the widening of the stem below the apex, it is better described as a frustum of a cone. These stem frusta at first grow mainly in width and then, only after several plastochrons, do they begin rapid elongation (Fig. 2.10) (Sunderland and Brown 1956, Lyndon 1987a). In extreme cases this initial growth in width has led to the concept of the primary thickening meristem, which causes the stem to widen very rapidly immediately below the apical meristem, as in some ferns (Wardlaw 1963) and also in palms (Tomlinson 1961) which do not have secondary thicken-

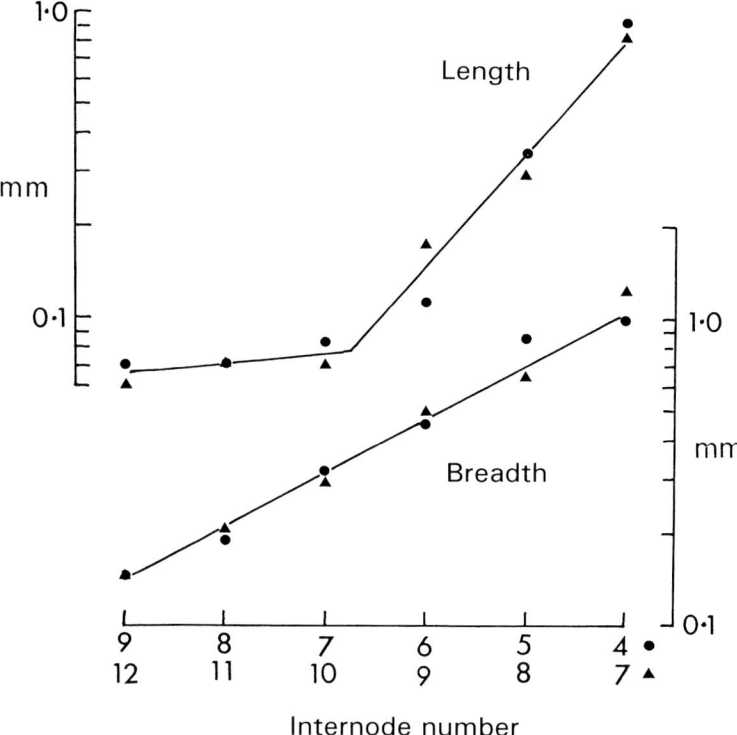

Figure 2.10 Internodes grow steadily (exponentially) in breadth from their inception at the shoot apex in *Silene*, but only begin to elongate rapidly after three plastochrons. From Lyndon (1987).

ing. The internodal cells in the youngest frusta sometimes also divide rapidly relative to their elongation rate so that they become evident as the pith-rib meristem.

2.7.   Formation of axillary buds

Axillary buds form in the axils of leaves and can first be recognized in longitudinal sections as the shell zone (Shah and Patel 1972), or arcuate zone, in the leaf axil. The axillary bud rather than forming in the leaf axil itself may often form at the base of the associated internode distal to the axil, and this may be why internodes and axillary buds tend to occur together, and why one does not occur without the other. In the flower, for instance, internodes are lacking and so are buds axillary to the floral organs. Clonal analysis in maize shows that axillary buds 'were usually clonally related to the internode and the leaf above the bud, rather than to the subtending leaf and internode' (McDaniel and Poethig 1988, p.19). This is what has also been suggested on anatomical grounds for maize (Sharman 1942). In *Cucurbita*, too, the buds form on the stem and are displaced laterally from the leaf midline (Snow 1965). Esau (1965, p.109) remarks that for many plants 'the term axillary is somewhat inaccurate because the buds generally arise on the stem . . . but become displaced closer to the leaf base, or even onto the leaf itself, by subsequent growth readjustments. . . . In the grasses, the lack of developmental relation between the bud and its subtending (axillant) leaf is particularly clear'. If axillary buds are associated with the internode above the subtending leaf, then they ought not to be detectable until the next leaf is formed, i.e. the axillary bud should be initiated at the same time as the internode and hence the leaf above. This is consistent with axillary buds commonly being initiated 'somewhat later than the leaves subtending them' (Esau 1965, p.110). The differentiation in cellular ultrastructure in the axil, which can be detected as soon as a leaf primordium is initiated (Lyndon and Robertson 1976), is therefore presumably related to the development only of the axil itself and not of the axillary bud.

2.8.   Origin of the shoot apical meristem

Shoot apex initials originate very early in embryo development. In *Cerastium* the cells of the shoot apex become distinguishable by their larger nucleoli but decreased cytoplasmic pyroninophila in the globular proembryo when it still consists of only about 150 cells (Mestre and Guignard 1967). Similarly in *Stellaria* apical meristematic cells became distinguishable as four cells at the tip of the globular proembryo shortly after the formation of the protoderm (Pritchard 1964).

Although the shoot apex initial cells become distinct in the globular stage of the embryo, the shoot apical meristem itself is not formed until later. In the pea, the apex structure is not fully developed until the embryo has made several primordia and the embryo is fully developed (Reeve 1948). The shoot apical meristem is initiated by periclinal divisions in the surface cells before the cotyledons appear. At an early stage of its development it shows two distinct outer cell layers, but since there are often periclinal divisions in both layers during early development of the apex, there is no clear tunica at this stage. In *Arabidopsis* there is a mutant, *shoot meristemless*, in which plants lack a shoot apical meristem

(Barton and Poethig 1993). The development of the mutants is the same as the wild-type until the torpedo stage of embryogenesis. Normally the meristem then forms between the cotyledons, but in the mutant the apparent initial cells develop no further. However, leaves are formed by adventitious primordia on the hypocotyl and also on the petioles of the leaves themselves, in the absence of a shoot apex (Barton and Poethig 1993).

The origin of the apical meristem has also been studied during development of the sporophyte from spores in the pteridophyte *Isoetes* (Michaux-Ferrière 1980). The apical meristem does not become distinct until the third leaf has been initiated. As leaves six to twelve are formed, the apical meristem grows to about 150μm in diameter, but no apical cell is apparent at this stage.

The formation of the apical meristem can therefore be later than the formation of the first leaves. In the formation of a new axis by the residual meristem of *Graptopetalum* the leaves seem to form first and the shoot apex later (Green 1994). In the *shoot meristemless* mutant of *Arabidopsis* (Barton and Poethig 1993), and in cultured explants of *Nasturtium* leaf axils (Selker and Lyndon 1996), the formation of leaves can take place in the absence of a shoot meristem. Primordium initiation therefore does not seem to depend on the presence of an existing shoot apical meristem. This is consistent with the apical meristem being primarily a source of cells and so providing a platform on which organ initiation can take place. A further indication that the shoot apex is primarily just a source of cells is that when sufficient of the apex is destroyed by a needle puncture, it forms primordia for a while but eventually stops when the existing supply of meristematic cells at the apex is used up (Sussex 1964). This is similar to what was found when the apical cell of a fern was punctured (Wardlaw 1949).

2.8.1.   *Origination of meristems* de novo *in culture*

Shoot meristems, vegetative and floral, readily form in cell and tissue cultures (Tran Thanh Van 1981). They can arise exogenously or endogenously, and from different cell layers, including the epidermis (Tran Thanh Van 1980). In callus on explants of cultured leaf discs of *Datura*, the shoot meristems were exogenous, with indole acetic acid (IAA) in the medium, or endogenous, with NAA (Iskander and Brossard-Chriqui 1980). In both cases their origin was similar: a meristematic front formed with an apparent tunica layer, and with lateral meristematic regions which initiated leaves.

2.8.2.   *Regeneration of meristems in split or damaged apices*

When the shoot apical meristem of *Lupinus* was divided by vertical cuts into six parts, some of these parts were each able to regenerate a new shoot apical meristem (Ball 1952a). This came about by cell divisions in the second tunica layer and in the corpus of the flank of the original apex (the uncut outer side). Cutting the original apex into eight parts did not allow the parts to regenerate: presumably there was a minimum cell and tissue mass necessary for regeneration.

2.9.   Function of the shoot apical meristem

The primary function of the meristem, at least the promeristem, is as a source of cells, or more properly cellular material, for the continued growth of the shoot. This is shown by the fact that the meristem is not necessary for leaf primordium formation; witness the initiation of leaves in the absence of a meristem (Barton and Poethig 1993, Selker and Lyndon 1996). Also, when the meristem is damaged and prevented from further growth, the formation of primordia continues until the existing meristematic cells are used up (Wardlaw 1949, Sussex 1963). The fate of these cells and the production of tissues and organs is determined only after the cells have left the shoot meristem. The only qualification to this is that by virtue of their position some meristem cells will be more likely to end up contributing to one part of the mature plant rather than another. The closer a cell happens to be to the summit of the meristem, the longer it is likely to remain as an initial before it is displaced down the apical dome by the cells that are the products of its own division, and of its neighbours. The number of initial layers in the shoot meristem tends to increase from gymnosperms to monocotyledons to dicotyledons, and this reflects some change in the factors controlling cell division orientations and hence presumably cellular growth rates at the summit of the meristem. It remains to be shown whether there are differences in cell structure or metabolism, or wall yielding properties that can be correlated with the number of initial layers.

3

Growth Rates Within the Shoot Apex

3.1. Growth rate of the whole shoot apex

The rate at which the shoot meristem grows determines the rate at which cellular material is available for further growth of the apical system, and so is one of the major determinants of the rates of shoot growth and leaf initiation, and may indeed be the major factor in limiting plant growth rate in extreme habitats (Körner and Menendez-Riedl 1989). Despite this being so, there is very little information about apical growth rates in relation to the growth of the plant as a whole. This is presumably because measuring growth rates and the control of growth at the whole plant or leaf level is relatively straightforward, whereas measuring growth rates of the generative cells at the apex is more difficult because of the minute size of the apex, which is also enclosed in the young leaves, making access to it difficult. However, if appropriate methods are used there is no reason why the growth rate of the shoot apex should not be measured and its role in the control of growth of the whole plant be clarified, especially for plants such as those growing in apparently favourable habitats in the tropics and where the main controls of growth rate seem to be internal.

43

Table 3.1. *Cell doubling times of shoot apical meristems from cell counts.*

Plant	Cell doubling time (h)	Reference
Tradescantia	96	Denne (1966b)
Agropyron	67	Smith and Rogan (1979)
Trifolium	64	Denne (1966a)
Secale	48	Sunderland (1961)
Lonicera	45	Edgar (1961)
Lupinus	32	Sunderland and Brown (1956)
Psium	28	Lyndon (1968a)
Silene	20	Miller and Lyndon (1976)
Vicia	8	Ball and Soma (1965)

3.1.1.   Methods of measurement

There are two basic methods for measuring growth rates of, and in, shoot apices. The first is by direct measurement of changes in cell number (Table 3.1) or apical volume. The use of cell numbers assumes that growth is accompanied by cell division and that mean cell volume remains essentially constant, or can be corrected for. The second is indirect, by measuring the rate of cell division or the length of the cell cycle and then using these to calculate growth rates.

3.1.1.1.   Counting cells. Mean cell doubling times in the *Lupinus* shoot apex were made directly by measuring the increase in cell number in successive growth units of the lupin apex, each unit representing a plastochron's growth (Sunderland and Brown 1956). On the assumption that the growth rate in the apical system as a whole did not change appreciably during the formation of the successive growth units, then the sequence of units in space represented a sequence of units in time, at intervals of a plastochron. The shoot tip was excised below the seventh youngest primordium and the shoot tissue was then cut into units each demarcated by a successive leaf primordium. The primordia themselves were also cut off and cell numbers in them and the remaining axial units were measured by squashing the tissue fragments to a layer one cell thick and then counting the number of cells. Volumes of the tissue fragments were measured by squashing them to a known thickness and then measuring the area of the squash.

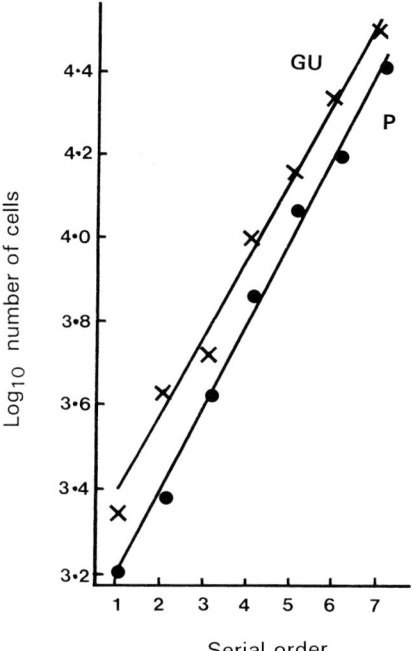

Figure 3.1 Exponential increase in cell number in successive growth units (stem axis plus primordium: GU, ×) and primordia (P, ●) of *Lupinus*, numbered in serial order, from the youngest (1) to the oldest (7) from the apex downwards. From Sunderland and Brown (1956).

The cell number in successive growth units increased exponentially (Fig. 3.1). The increase in cell number in the primordia (from their initiation) was also exponential with a doubling time of 1.48 plastochrons, and since the plastochron was about 2 days, this implies a cell doubling time of about 70h. The cell doubling time of the shoot apical dome was also calculated. In the course of a plastochron (i.e. 2 days) a dome of 3500 cells produced a growth unit (comprising the dome plus the first growth unit of primordium plus axis) of about 5800 cells. Assuming exponential growth in the apical dome this gives about 66h for the cell doubling time of the apex, which was essentially the same as that for the youngest primordia.

The same principle for measuring the cell doubling time of the apex can be used with sections. By counting all the cells in serial sections of the apical dome, and in the apical dome plus the first growth unit, and by taking samples at intervals during a single plastochron (Lyndon 1968a), or simply at the beginning and end of a plastochron, the increase in cell

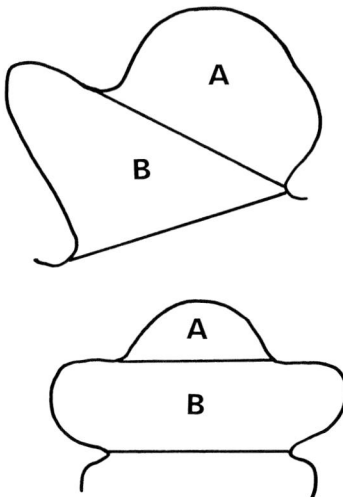

Figure 3.2 An estimate of apical growth rate can be obtained from the increase in cell number or tissue area in a L.S. of an apex. The upper apex is forming primordia one at a time on alternate sides of the apex; the lower apex is forming a pair of primordia each plastochron. A line joining the youngest leaf axils demarcates the apical region A from the product of one plastochron's growth, B. A gives rise to A + B in the course of one plastochron.

number during a plastochron can be found and then, assuming exponential growth:

cell doubling time = (log2/increase in log cell number)
× length of plastochron (h).

For plants with opposite or distichous leaf arrangements, the procedure can be simplified by the use of only the median section, since the cell number in the median section is closely proportional to total cell number in all the serial sections of the apex (Miller and Lyndon 1976). This also means that if samples are restricted it is possible to get an estimate of cell doubling time even if only a single representative section is available and the length of the plastochron is known. The apical dome gives rise in one plastochron to a new apical dome plus the primordium plus associated axis. So, a line across the median section of an apex joining the axils of the second and third youngest primordia isolates all the tissue above as being the product of one plastochron's growth, and another line joining the axils of the first and second youngest primordia demarcates the apical dome, which has produced this tissue in the course of one plastochron (Fig. 3.2). The cells in each of these demarcated regions can be counted. Assuming growth in this region of the apex to be exponential, then where A is cell number in the apical dome, and $A + B$ is cell number in the apical dome plus primordium and associated axis, then the relative growth rate (RGR) (h^{-1}) is:

$$RGR = [\log_e(A + B) - \log_e A]/\text{plastochron (h)}$$

and

$$\text{cell doubling time (h)} = \log_e 2/RGR.$$

In *Silene coeli-rosa* the cell doubling time for the shoot apex of vegetative plants was 20–24h (Miller and Lyndon 1976). The use of only the median, instead of all, sections led to an error of $\leqslant 8\%$ on average.

Useful estimates of cell doubling time can be obtained in this way, especially where material is limited, so long as the rate of leaf initiation can be measured, and with the assumption that the phyllochron approximates to the plastochron (the phyllochron being the interval between the visible emergence of successive leaves). Although this method has hardly been used, it would seem to be potentially very useful for surveys of apical growth rates in the field where many species and individuals could provide measurements. It would be of great interest to find out what apical growth rates are in extreme habitats (Körner and Menendez-Riedl 1989), and in tropical habitats where plant growth rate may not be restricted by environmental variables but might be constrained by the rates of growth of the shoot apical meristem.

An increase in cell number as a function of real time can be found by sampling apices at intervals of growth during at least one plastochron. However, for accurate, detailed data it is important to take samples as a function of developmental stage not of chronological age (Denne 1966a, b, Lyndon 1968a). If samples are taken at time intervals, irrespective of their developmental stage, the normal variation in rate of development from plant to plant means that a small range of different developmental stages tends to get grouped in a single sample, so that data obtained represent an average of the developmental stages in the sample. This is the case for the lupin data of Sunderland and Brown (1956) (see above) so they therefore had to assume that the values obtained represented mid-plastochron values. However, if a developmental time scale is used to sample the plants, such as the plastochron index (Erickson and Michelini 1957) or a similar developmental scale of plastochron stage (Lyndon 1968a), then precise developmental stages within a plastochron can be sampled and the mean times for these to be achieved can be worked out.

In the pea apex, the increase in cell number was measured during the course of a single plastochron of 46h, and gave a mean cell doubling time for the apex of 28h (Lyndon 1968a). In *Silene* apices, the volume doubling times for the whole shoot apex were converted to cell doubling times (Lyndon 1979a) and the 26h for vegetative meristems growing at 20°C was close to the 20–24h obtained from counting cells directly (Miller and Lyndon 1976).

These measurements depend on mean cell volume being similar throughout the apex. If it is not, for example, the cells in the pith–rib meristem perhaps tending to enlarge, then this must be corrected for. Although mean cell volume may often remain more or less constant over the period of measurement, in some apices it may increase appreciably in the course of a single plastochron. In *Trifolium*, mean cell volume in median sections of the shoot apex increased from 364 to 464μm^3 during one plastochron. Correcting for this, the increase in apical volume allowed the calculation that the cell number increased, exponentially, from about 2000 to about 5000 cells over a plastochron of 86h; the cell doubling time was therefore 64h (Denne 1966a). In the grass *Agropyron*, using median sections as the basis for volume calculations in successive plastochrons, it was shown that cell volume tended to decrease in successive plastochrons so that volume doubling time was longer than the cell doubling time, which was 67h in the young, vegetative plant (Smith and Rogan 1979).

3.1.1.2.   Plastochron ratio. Doubling times can also be obtained from measurements of the plastochron ratio. The growth of the apex can be followed at intervals of time because the apex produces natural markers of its growth, the leaf primordia, at regular intervals, each interval being a plastochron. Leaf primordia are formed on the flanks or periphery of the apex and always at some minimum distance from the apical summit or centre. The ratio of the radial distances of successive primordia from the apical centre is the plastochron ratio, r (Richards 1951) (see Fig. 8.4). A constant plastochron ratio for successively older pairs of primordia at the apex implies exponential radial growth of the apex, and so the relative radial growth rate per plastochron is given by $\log_e r$. The cross sectional area relative growth rate is therefore $2\log_e r$ and, if it is assumed that the vertical component of growth in the apex is also exponential, and the same as the radial rate, then the volume relative growth rate of the apex is $3\log_e r$ and $\log_e 2/3\log_e r$ is the volume doubling time in plastochrons.

Table 3.2. *Volume doubling times of shoot apical meristems derived from the plastochron ratio.*

Plant	Volume doubling time (h)	Reference
Eleais	1150	Rees (1964)
Dryopteris	480	Richards (1951)
Lupinus	120	Richards (1951)
Pinus	120	Cannell (1976)
Picea	78	Romberger and Gregory (1977)
Picea	51–69	Cannell (1978)
Chrysanthemum	70	Schwabe (1971)
Chrysanthemum	50	Berg and Cutter (1969)
Triticum	47	Evans and Berg (1971)

When the length of the plastochron is known the volume doubling time in hours can be found (Table 3.2). Assuming all the cells in the apex to be dividing and the mean cell volume to be essentially constant, this is equal to the mean cell doubling time.

A refinement of this technique was used very effectively to measure cell doubling times in *Pinus* and *Picea* shoot apices (Cannell 1976, 1978). Each shoot apex was fixed and exposed by dissecting away the leaves and cataphylls that obscured it. A tracing of the exact phyllotactic arrangement of the primordia was made and it was then possible to match each tracing accurately to one of a series of more than 200 computer-generated patterns, representing increments of 0.002 in the plastochron index (Fig. 3.3). The shortest volume (and cell) doubling times of apices of *Pinus* were about 120h (Cannell 1976). Using the same technique, the shortest doubling times in *Picea* shoot apices were shown to be about 50h (Cannell 1978).

3.1.1.3. Colchicine and labelling methods. Colchicine has been used to inhibit mitosis by preventing the formation of the spindle so that cells accumulate at metaphase. The mean rate of cell division in the whole apex can then be measured, and the growth rate of the shoot apex can then be found (Lyndon 1970a). The cell cycle in shoot apices can be measured by pulse-labelling with radioactive thymidine, but cell cycle times can be translated into apical growth rates only where all the apical cells are cycling cells (Miller and Lyndon 1975, 1976, Francis and Lyndon 1978b). Where some of the apical cells may not be cycling

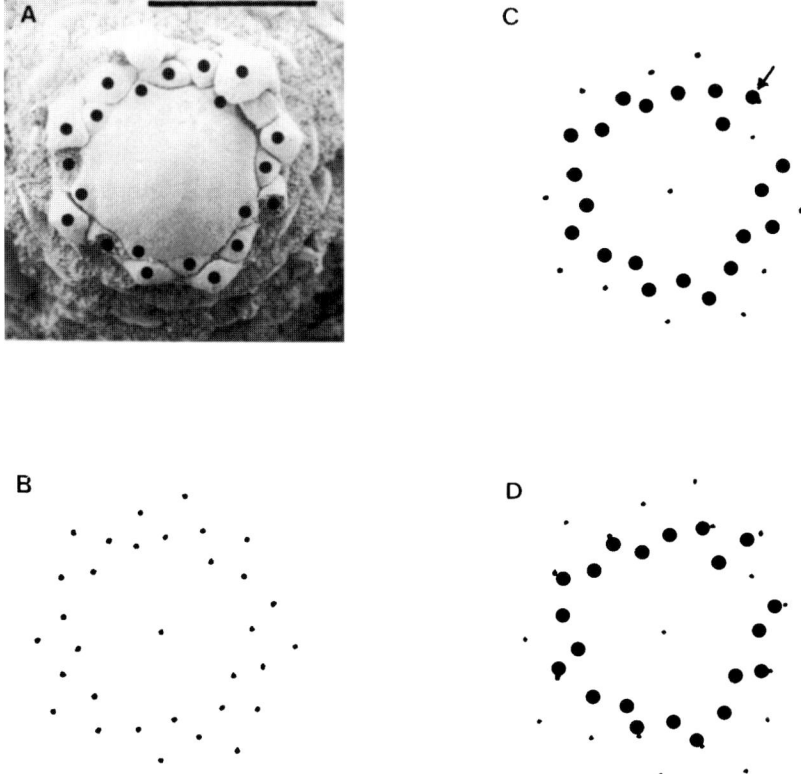

Figure 3.3 Plastochron ratio measured in apices of *Pinus*. On a photograph of the apex (bar = 0.5mm) the positions of the primordia are marked to give the pattern (A). A pattern of points with divergence angle of 137.5° and a plastochron ratio of 1.018 is generated by computer (B). When the pattern of primordia (A) is superimposed on the computer-generated pattern (B), the fit (C) is good, except for the most mature primordium (arrow). When the primordium pattern (A) is superimposed on a similar plot but with a slightly different plastochron ratio, 1.022, the fit is less good (D). In this way the plastochron ratio of *Pinus* apices could be determined to within ±0.002. From Cannell (1976).

(Gonthier, Jacqmard and Bernier 1985) then labelling data cannot be used to calculate overall growth rates unless the proportion of cycling cells is known precisely.

There needs to be caution in accepting the absolute values for cell doubling times obtained with these methods because of the possibility that the experimental treatment itself inhibits growth. In *Pisum*, colchicine inhibited growth by about 50%, as shown by comparing the cell doubling time obtained with that calculated from the rate of increase in cell number in untreated apices. The cell doubling times obtained with colchicine were therefore corrected to take account of this (Lyndon 1970a). Cell doubling times obtained from cell counts in untreated apices and apices labelled with ^{3}[H]thymidine were the same in *Silene* (Miller and Lyndon 1975, 1976), even though the apices had to be partially dissected before treatment, and so the treatment here seemed not to affect apical growth. Excising some of the youngest primordia before application of the labelled solution was found to be the most effective method of labelling the shoot apex (Bernier and Bronchart 1963). Values for cell doubling times on the flanks of the *Chrysanthemum* apex obtained by labelling (Table 3.4) were comparable with those measured from the plastochron index (Table 3.2). It is not clear whether some of the very long cell doubling times (of several hundred hours) reported for potato stolons, and the faster doubling time of the central zone than the flank cells, might have been because of inhibitory effects of the colchicine (Clowes and MacDonald 1987). The rather long cell doubling times, of up to 1000h, measured with colchicine in *Dactylis* were, however, confirmed when volume doubling times were measured from sections by the growth unit method (see Section 3.1.1.1. and Fig. 3.2) (D. Francis, unpublished data). Although colchicine inhibited the rate of cell division in *Pisum* apices, it was argued that, because the pattern of colchicine metaphases did not change with time of exposure to the colchicine, the relative rates of division so measured were valid even though the absolute values required correction (Lyndon 1970a).

3.1.2. Are mitotic indices a valid measure of division rates?

Now that measurements of division rates by direct methods (colchicine or labelling) are available it is possible to examine the usefulness or otherwise of the mitotic index as a measure of growth and division rates. If the absolute length of time spent in mitosis is constant and if the cell cycle as a whole lengthens (and division rate therefore decreases), then in an asynchronous meristem the mitotic index will be inversely proportional to the length of the cell cycle and proportional to division rate. If, however, the time spent in mitosis varies then there will be no clear relationship between mitotic index and division rates or the cell cycle.

In the moss *Hookeria*, the mitotic index was inversely proportional to the length of the cell cycle (measured by continuous labelling) and so was

Table 3.3. *Length of mitosis in shoot apices.*

Plant	Mitosis (h)	Reference
Mosses		
Polytrichum	3.6–3.8	Hallet (1977)
Leucobryum	1.4	Hallet (1977)
Hookeria	2.0	Hallet (1977)
Azolla	2.8	Gifford and Polito (1981)
Ceratopteris	7.0	Polito (1979)
Angiosperms		
Trifolium	3.3–4.2	Denne (1966a)
Pisum	1	Lyndon (1973)
Datura	1.5–2.2	Corson (1969)
Lonicera	5.1	Edgar (1961)
Chrysanthemum, vegetative	3.2	Nougarède and Rembur (1978)
Helianthus, veg 8 day	2.0–3.1	Marc and Palmer (1984)
Helianthus, fl 16 day	1.8–3.1	Marc and Palmer (1984)
Triticum, vegetative	3.1	Griffiths (1981)
Triticum, flowering	1.7	Griffiths (1981)
Isoetes	3	Michaux (1969)
Silene	0.6–2.6	Francis and Lyndon (1978a)
		Taylor and Francis (1989)
Sinapis	1.2–3.6	Bernier et al. (1993)

proportional to the rate of division (Hallet 1976). The length of mitosis was therefore essentially constant at 2.6–2.9h, for all cells where it could be measured. This is similar to the values for higher plants (Table 3.3). Assuming that in *Hookeria* mitosis also takes about 2.8h in the apical cell, then the low mitotic index of 1.1% for the apical cell implies a cell cycle of about 260h. This compares with the minimum cycle time of 105h indicated by continuous labelling, with a 95% probability that it is longer than this. Such very low mitotic indices for the apical cell do not imply that it is not dividing, only that the cell cycle is very long. Low mitotic indices therefore cannot be used as evidence for the apical cell not being an active initial. The cell cycle of the apical cell can be deduced in those mosses with three rows of leaves (corresponding to the three basiscopic faces of the apical cell) by the rate of leaf initiation since one leaf is produced by each merophyte and each merophyte results ultimately from one cell division in the apical cell (Hallet 1977).

Where direct comparisons have been made of mitotic index and the rate of division in higher plants, it has also been shown that they are proportional to each other, implying that the absolute time spent in mitosis is the same for cells with different cell cycles but in the same meristem (*Trifolium*, Denne 1966a; *Datura*, Corson 1969; *Sinapis*, Bodson 1975; *Chrysanthemum*, Nougarède, Rembur and Saint-Côme 1987, Nougarède et al. 1990). This conclusion is in general supported by many data showing that the time spent in mitosis tends to be constant in a meristem even though there are different cell cycle lengths of the component cells and that these may alter as a result of experimental treatment (see Tables 4.1, 4.4, and Figs 4.2, 4.3, 4.6). It seems, then, that most of the work in which relative rates of division within a given species were inferred from differences in mitotic index is probably valid.

However, there may be situations in which the mitotic index is not a valid indicator of the division rate. For example, in pea roots a decrease of temperature from 25 to 15°C doubled the length of mitosis but hardly altered the mitotic index because it also almost doubled the length of the cell cycle (Brown 1951). While it may therefore be acceptable to use mitotic index as an indicator of division rate in comparable meristems under similar conditions, it may be unwise to use it in other situations. It is not known whether mitosis is constant in length between different types of meristems in the same plant, nor whether there may be environmental conditions or experimental treatments that may alter the length of mitosis and so alter the relationship between mitotic index and division rate. Mitotic index cannot be used to provide an indicator of relative rates of division between different species. For this to be possible, mitosis would have to take the same time in all species and under all growing conditions, an unlikely proposition. Even in the same species we should beware that extreme environmental conditions may result in mitosis taking longer. Indeed, a good way of demonstrating mitosis in root tips is to cool them, since cell division stops, with the cells arrested in mitosis, in which case the persistence of mitotic figures and a positive mitotic index in fact reflects a zero rate of division.

3.1.3.   How fast does the shoot meristem grow?
Doubling times of apical meristems

The growth of the shoot apex as a whole appears to be exponential, as would be expected of a system in which all or a constant proportion of the cells are meristematic and cycling. An exponential increase in cell number in the shoot apex in successive plastochrons has been demonstrated in the lupin apex (Sunderland and Brown 1956, 1976) and within a single plastochron, the measurements being restricted to the apical dome and the youngest primordium (Edgar 1961, Denne 1966a, Lyndon 1968a, 1977b).

The cell doubling times (sometimes referred to as the mean cell generation times) for vegetative shoot meristems are mostly of the order of 1–4 days but range from 8 to 1150h (Tables 3.1 and 3.2). This latter, very long, value (48 days) was for the oil palm (*Eleais*) growing in West Africa in what would seem to be a favourable environment (Rees 1964). When *Eleais* embryos were grown in culture under conditions for inducing callus, although the cell doubling time was shorter it was still relatively long, about 185h, or of the order of a week (Pritchard 1977) and so a long cell doubling time seems characteristic of *Eleais*. Cell doubling time may also be long in the shoot apices of Brazilian cerrado plants, which show slow growth even when grown under apparently optimal controlled conditions (Felippe and Dale 1990). Cell doubling times in the apex of these and many other plants growing slowly in apparently favourable conditions are not known but raise the question of what the controls on apical growth rate really are. In all plants these controls must be ultimately intrinsic to the species and may depend on the nucleotype or other innate controls of growth rate and the cell cycle of which we are so far unaware. Cavalier-Smith (1985) suggests that the control of growth rate is essentially by the control of the rate of increase of cell volume, which in turn determines cell cycle length.

How do these values compare with those for the roots of the same plants? Probably root cells divide faster (cycle faster) than shoot cells. Clowes (1976, p.263) notes that 'shoot apices . . . seem always to have lower rates [of mitosis than root apices] in the few plants that have been investigated'. Values for the cell cycle in roots tend to be shorter (5–20h), than in vegetative shoot meristems (1–3 days). Direct data comparing the root and shoot apices of the same plant at the same temperature and at comparable stages of development are available only for *Dactylis*, in which the cell cycle at 20°C was 28–38h in the shoot apex and 9.5–12h in the root apex (D. Francis, unpublished data). Nevertheless, it seems probable that the average cell cycle in the vegetative shoot meristem is longer than in the root, because the root meristem extends only from about 0.5–1 or 1.5mm from the root tip and so is relatively small, but has to generate all the cells used for growth of that root axis. In contrast, meristematic activity in the shoot meristem persists into the young leaves and stem so that perhaps the majority of the cells in the apical bud are cycling. With such a large capital of so many cycling cells the cell cycle in the shoot can be longer than in the root and yet still be quite adequate to generate large numbers of cells rapidly.

One comparison that can be made is of the growth of the cells, and the nucleus and nucleolus in the root and shoot meristems of the pea. Mean cell volume for cells with prophase mitotic figures in the root meristem was $1740\mu m^3$ (Lyndon 1967; cultivar Meteor) but about $800\mu m^3$ in the shoot apex (Lyndon 1970c; cv. Lincoln). This corresponded to nuclear volumes (of early prophase nuclei) of $460\mu m^3$ in the root and $310\mu m^3$ in the shoot meristems (Lyndon 1968b); and ratios of 2.5 for cell volume and 1.5 for nuclei. Also, a ratio of root/shoot nuclear volumes of 2.1 can be calculated for nuclei of proliferating cells in *Helianthus* (Langenauer, Davis and Webster 1974). Because the meristematic cells tend to be smaller in the shoot than in the root, and because the cell cycle tends to be longer, then the rate of cell growth will be less in the shoot meristem than in the root meristem.

The nucleolus disappears at mitosis and so is reformed in each cell cycle. The nucleolus grows throughout interphase (Lyndon 1977a) and its final pre-prophase size in the shoot cells of pea, $8\mu m^3$, is only one-third of that, $25\mu m^3$, in the root cells (Lyndon 1968b). If the cell doubling time is longer in the shoot meristem than in the root meristem, this implies a much slower rate of nucleolar growth, and perhaps of synthesis of rRNA, in the shoot than in the root. In the flowering apex, the rate of rRNA synthesis probably increases and may be linked to an increased availability of ribosomal DNA for transcription (see Chapter 9).

3.2. Heterogeneity of growth rates within the apex

Shoot apical meristems as seen in sections are not homogeneous. Subapical cells are different in size and shape from apical cells; in gymnosperms there is often a group of so-called central mother cells, and in angiosperm apices tunica and corpus can be distinguished. Different regions of the apex also react differently to stains: often there is a central region which stains less than the peripheral regions (see Chapter 5). Of especial interest has been the observation that mitotic figures are often less frequent at the summit of the apex and more frequent on the flanks, suggesting differences in cell division rates (Nougarède 1967). Together with the fact that leaf primordia are initiated on the flanks of the apex, this all suggests that there are gradients of cell division and growth rate within the apex.

Differences in growth rates within the apex are most readily measured from the rates of cell division or from the cell cycle. If there are appreciable changes in cell volume during the measurements then this has to be allowed for in converting division rates to tissue growth rates. Measurements can also be made of rates and directions of surface displacement, from which likely internal growth rates can be inferred.

By using an inhibitor that is specific for halting mitosis, the rate at which cells accumulate in mitosis can be measured and this would equal the rate at which cells enter mitosis in the untreated plant. The rate of cell division can then be calculated and the reciprocal value gives the mean cell doubling time of the apex, assuming that all cells are dividing. Colchicine complexes with the tubulin protein which forms the spindle and so prevents spindle formation and inhibits the cells at metaphase. The rate of accumulation of metaphases gives the rate of division. However, colchicine can also inhibit growth. In *Pisum* the cell doubling time calculated from the fastest rate of accumulation of colchicine metaphases was only about half that expected from measurements of doubling time of cell number (Lyndon 1970a). However, since the distribution of C metaphases throughout the apex remained the same at all times during colchicine treatment, it seemed that the rate of cell division was inhibited to an equal extent in all parts of the apex. Therefore the colchicine data could be justifiably used to compare differences in division rates in different parts of the apex (Lyndon 1970a). A major advantage of using colchicine is that the position of every C metaphase, i.e. every potential division, can be noted so that maps can be made of the distribution and frequency of divisions in all regions of the apex (Lyndon 1970a) (see below and Fig. 3.4).

If it can be assumed that all the cells of the shoot apex are cycling cells, then the doubling time can be measured as the mean length of the cell cycle measured by the pulse-labelling method (PLM) or by the double-labelling method devised for measurements in the shoot apex where there are usually few visible mitotic figures (Miller and Lyndon 1975, Francis and Lyndon 1978b) (see Chapter 4 for details).

3.2.1. Growth gradient: higher plants

It has often been noticed that the frequency of mitotic figures at the summit of the shoot meristem is less than on its flanks (Nougarède 1967). Assuming that the mitotic index is proportional to growth rate

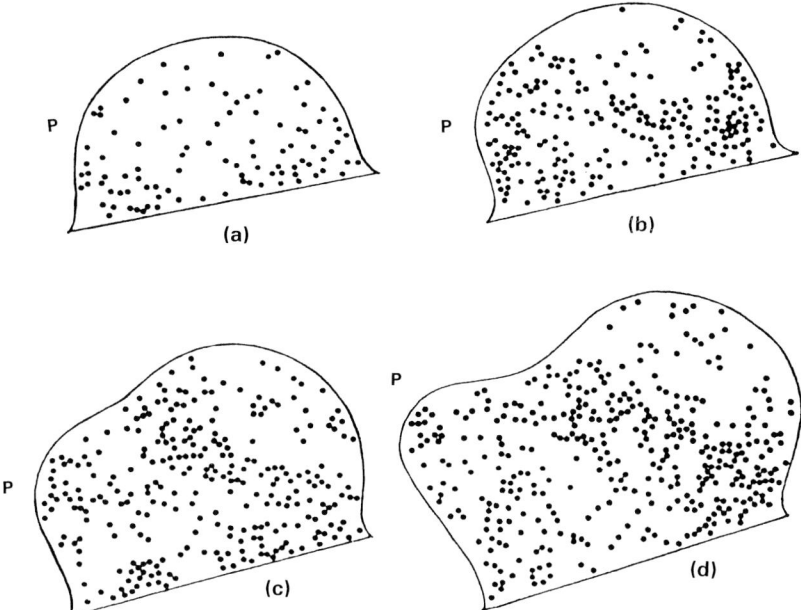

Figure 3.4 Distribution of cell division, as revealed by the occurrence of colchicine metaphases, in pea (*Pisum*) apices (L.S.) at four stages of a single plastochron. At the beginning of the plastochron (a) the primordium (P) is just starting to emerge, and cell division is of relatively low frequency in the apical dome but higher frequency at the base of the dome. As the plastochron progresses (b–d) the primordium grows out and a band of more frequent divisions becomes apparent at the base of the apical dome. From Lyndon (1970a).

(the higher the mitotic index, the higher the growth rate, and vice versa), then this leads to the conclusion that the cells at the summit of the shoot apex are growing and dividing more slowly than the cells on its flanks. This led to the hypothesis of an 'anneau initial' or 'initiating ring' on the flanks of the apex and a 'méristème d'attente' or 'waiting meristem' at the summit (Nougarède 1967). The waiting meristem was believed to become active only on the transition to flowering, but this is now known not to be so (see Chapter 2, Section 2.2.2). However, it was not until direct measurements of division and growth rates were made that it could be clearly shown that cells in different regions of the shoot apex were indeed growing at different rates. Since mean cell volume in the apex appears to remain essentially the same over the periods of measurement, it may be assumed that rates of cell division can be used as measures of growth rate and that, vice versa, measurements of apex volume doubling times can be used to estimate the mean cell cycle length.

Table 3.4. *Cell doubling time (mean cell generation time) at the summit and on the flanks of vegetative shoot apical meristems of angiosperms.*

Plant	Cell doubling time (h)		References
	Summit*	Flanks**	
Trifolium	108	69	Denne (1966a)
Pisum	69	28	Lyndon (1970a)
Pisum (main apex)	49	31	Nougarède and Rondet (1976)
Pisum (axillary bud, inhibited)	127	65	Nougarède and Rondet (1976)
Pisum (axillary bud, released)	40	33	Nougarède and Rondet (1976)
Oryza	86	11	Rolinson (1976)
Rudbeckia	> 40	30	Jacqmard (1970)
Solanum	117	74	Leshem and Clowes (1972)
Datura	76	36	Corson (1969)
Coleus	250	130	Saint-Côme (1971)
Sinapis	288	157	Bodson (1975)
Chrysanthemum[a]	144	50	Nougarède et al. (1990)
Chrysanthemum[b]	102	32	Nougarède et al. (1990)
Chrysanthemum	139	48	Rembur and Nougarède (1977)
Chrysanthemum	140	54	Nougarède and Rembur (1977)
Helianthus	83	37	Marc and Palmer (1984)

*Or central zone; **Or peripheral zone.
[a]Photon flux = 70μmol/m^2.
[b]Photon flux = 200μmol/m^2.

The first direct measurements of the rates of cell division in different regions of the apex were made in *Trifolium* by the method of the accumulation of colchicine metaphases (Denne 1966a). This was soon followed by similar measurements on other plants. Later, radioactive labelling, either double-labelling or pulse-labelling, was used to get detailed information of the lengths of the various phases of the cell cycle. All these measurements (except for a Swedish population of *Dactylis*; D. Francis, unpublished data) are consistent in showing that in the adult vegetative apex the cells of the apical summit are dividing

Table 3.5. *Shortening of the cell doubling time (mean cell generation time) at the summit and on the flanks of shoot apical meristems on transition to flowering.*

	Cell doubling time (h)			
Plant	Summit	Flanks	Ratio (S/F)	Reference
Chrysanthemum (vegetative)	140	54	2.59	Nougarède and Rembur (1977)
Chrysanthemum (prefloral)	54	47	1.15	Nougarède and Rembur (1977)
Datura (vegetative)	76	36	2.11	Corson (1969)
Datura (floral)	46	26	1.85	Corson (1969)
Coleus (vegetative)	250	130	1.92	Saint-Côme (1971)
Coleus (floral)	130	130	1.00	Saint-Côme (1971)
Sinapis (vegetative)	288	157	1.83	Bodson (1975)
Sinapis (floral)	35	25	1.40	Bodson (1975)
Helianthus (vegetative 8 day)	83	37	2.24	Marc and Palmer (1984)
Helianthus (floral 16 day)	33	39	0.85	Marc and Palmer (1984)

more slowly and have a longer doubling time than the cells on the apical flanks (Table 3.4). This has now been observed directly in the apex of *Anagallis* where, over the course of two plastochrons (Fig. 2.2), the more peripheral cells have approximately twice the area relative growth of the cells at the summit of the apical dome (Green, Havelange and Bernier 1991). On the transition to flowering, the growth rate of the apex often increases (see Chapter 9) and the cell cycle shortens relatively more in the central zone than on the flanks of the apex, so that the steepness of the growth gradient is reduced or disappears altogether, as in *Coleus* and *Helianthus* (Table 3.5).

The heterogeneity of cell division rates has been visualized in pea apices by making three-dimensional reconstructions of colchicine-treated apices so that the detailed distribution of dividing cells can be followed through the plastochron (Fig. 3.4). In *Pisum* the whole of the apical dome is relatively slowly growing and, as the plastochron progresses, a plate of more rapidly dividing cells becomes visible at the base of the apical dome, with a possible locus of fastest division just below the surface of the I_1 region where the next primordium will emerge (Fig. 3.4 and see Fig. 4.1).

The plate of dividing cells corresponds in position and time of appearance to the cambial-like layer that has been noted in some meristems (Philipson 1954). The position of the putative procambial strands is also first marked out by clusters of faster dividing cells (Lyndon 1970a).

The difference in cell cycle lengths between the summit and the flanks in vegetative meristems is typically about two-fold, sometimes more. Labelling experiments with *Helianthus* and *Nicotiana* showed no detectable growth in the summit cells of these meristems within several days (Steeves et al. 1969, Sussex and Rosenthal 1973, Langenauer, Davis and Webster 1974). Since cell sizes in the shoot apex tend to remain approximately constant, at least over periods of the order of a cell cycle, then growth rates and division rates tend to vary together and one can be taken as a measure of the other. A gradient of an increasing rate of growth and cell division from the apical summit or centre is therefore the rule for adult vegetative meristems initiating leaves on their flanks, the highest growth and division rate being where the primordia are formed. Growth rates may be maintained or even increased in the internodes below the apex, as in *Lupinus* (Sunderland and Brown 1956).

When the apex is disturbed, for instance by excision and culture, cell division in the central zone of the meristem is stimulated but subsides again, and the cells become quiescent once steady growth of the apex is resumed (Sussex and Rosenthal 1973, Davis, Rennies and Steeves 1979).

3.2.2.   Growth gradient: plants with apical cells

Mosses and pteridophytes, with an apical cell, also show a gradient in the rate of cell division from least in the apical cell to faster division in the derivative cells and so, in this respect, they are like vegetative meristems in higher plants.

In the mosses cell cycles differ in different regions of the apex in much the same way as in higher plants. A detailed review of the different methods used to determine the cell cycle of apical cells is given by Paolillo (1984). The cell cycle on the flanks of the apex (i.e. in the derivatives of the apical cell) and in the young leaf primordia, tends to be two or three times shorter than in the apical cell (Table 3.6). However, in the two water ferns studied (*Azolla* and *Ceratopteris*) the cell cycle in the apical and other cells was very similar (Table 3.6). Is this because they are aquatic plants, or because of the slender shape of their apices? In

Table 3.6. *Cell doubling times in apical and derivative cells in mosses and pteridophytes.*

Plant	Cell doubling times (h)		Reference
	Apical cell	Derivatives	
Mosses			
Polytrichum (January)	186	80	Hallet (1974)
Polytrichum (March)	c.400	110	Hallet (1974)
Polytrichum (May)	192	70	Hallet (1974)
Hookeria	> 105	90	Hallet (1976)
Pteridophytes			
Isoetes (young)	21	20	Michaux-Ferrière (1980)
Isoetes (adult)	> 53	36	Michaux (1969)
Polypodium (young)	62	61	Michaux-Ferrière (1981a)
Polypodium (adult)	144	78	Michaux-Ferrière (1981a)
Pteris	72	60	Michaux (1971)
Azolla	28	27	Gifford and Polito (1981)
Ceratopteris	69	75	Polito (1979)

Oryza (rice), a higher plant with an elongated apex, there is a characteristic growth gradient (Table 3.4). In *Equisetum* the mitotic index (7.0%) was higher in the derivative cells below the apex than in the apical cell and its immediate derivatives (3.9%) (Gifford and Kurth 1983). Assuming the mitotic index to be proportional to division rates then the *Equisetum* apex, which is also elongated, seems to resemble that of the mosses and higher plants in having a longer cell cycle in the apical cell than on the flanks of the apex, and so has a gradient of increasing growth rate from the summit of the apex to the flanks.

In the pteridophytes, the cell cycle of the *Pteris* apical cell (72h) is longer than in its immediate derivatives (60h) (Table 3.6) and is twice as long as the cell cycle in the cells a little further from the apical cell (36h) (Michaux 1971), so that growth increases away from the apical cell. However, there appears to be no growth gradient in the juvenile meristem of *Isoetes* (Table 3.6). Unlike the adult meristem it shows no zonation with respect to protein or DNA distribution or ultrastructure; also the ratio of nucleus to cell volume is the same throughout the meristem and so is the mitotic index (*c.*7%) (Michaux-Ferrière 1980). The cell cycle, measured from continuous labelling, is also about 20h throughout, with G_2-phase < 4h and M-phase *c.*1.4h. Since the initial labelling index was 10–18% this implies S-phase is 2–3.6h, and G_1 *c.*11–13h (and therefore the major part of the cell cycle), and these values are similar throughout the apex. Whether in higher plants there is also a lack of growth gradient in apices of young plants is not known.

3.3.   Growth rates in relation to meristem shape and cellular structure

The cells in a meristem are not randomly arranged. They lie in files, and may not be isodiametric but elongated along a preferred direction. The pattern of cells in a meristem is often characteristic of a species or taxon (Hejnowicz 1955). The most obvious pattern in most angiosperm shoot apices is the distinct tunica. The pattern of cells in the meristem arises because the orientations of the new cell walls in the dividing cells tend to be normal to the principal directions of growth (Hejnowicz 1982, 1984). To produce the cell pattern seen in the angiosperm apical dome, including the presence of a tunica, an increase in growth rate from the summit down the axis and surface of the apical dome is required (Hejnowicz 1955, Hejnowicz, Nakielski and Hejnowicz 1984a), combined with an increasing anisotropy of surface growth, becoming increasingly longitudinal with distance from the apical summit (Hejnowicz, Nakielski and Hejnowicz 1984b) (Figs 3.5c, 3.6). Similarly, the only growth pattern that is consistent with a gradient of growth rate from minimum at the summit to maximum on the flanks of the apex, as observed in most shoot apices, is also one in which growth is anisotropic with the longitudinal component predominating and increasing from the summit to the flanks. A gradient of growth rate from minimum at the summit to maximum on the flanks of the apex is not consistent with either isotropic growth (Fig. 3.5a) or anisotropic but mainly transverse growth (Fig. 3.5b); in both cases growth rate would have to be maximal at the summit and least at the base of the apical dome, contrary to what is found in shoot apices.

Hejnowicz (1982) has shown how internal rates of growth could be inferred if rates of displacement of points over the surface of the apical dome and the shape of the apical dome are known. Although the displacement of points over the apical surface has been followed (Loiseau 1962, Soma and Ball 1963), the actual rates of displacement have not yet been measured. Marks placed more or less centrally on the apical meristem of *Impatiens* became displaced off the meristem in about four-and-a-half plastochrons on average (Loiseau 1962). The marks stayed for a long time without any apparent movement and then slid slowly towards the periphery, elongating radially and accelerating as they approached the

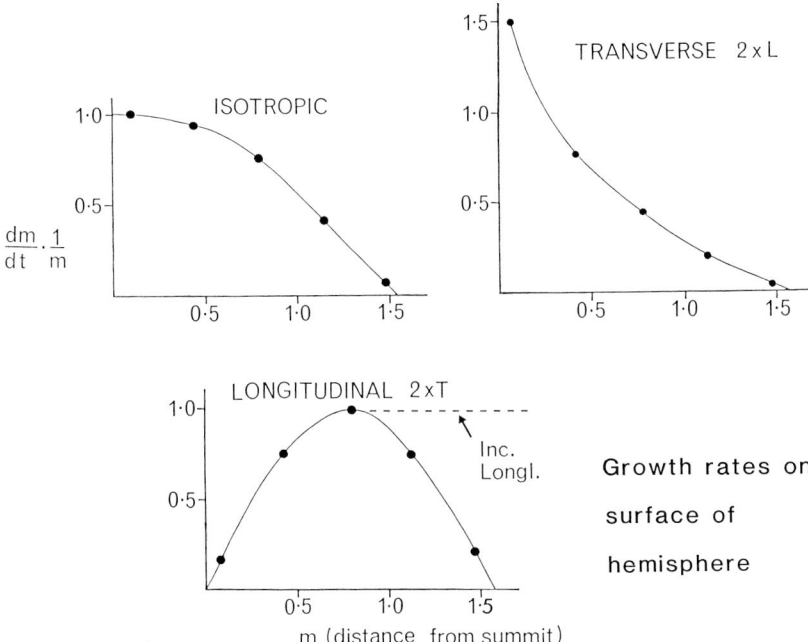

Figure 3.5. Growth rate down the apical dome in relation to anisotropy of surface growth. Isotropic growth and anisotropic growth with a predominantly transverse component (2× longitudinal component) would each require a decrease in growth rate from maximum at the summit of the dome to zero on the flanks, which is not what is found in apices. Anisotropic growth with a predominantly longitudinal component (2× transverse component) would require minimal growth at the apical summit increasing to a maximum on the flanks. This is consistent with what is observed. If the high growth rate on the flanks is maintained to the base of the apical dome, and becomes increasingly longitudinal (Inc. Longl.), this corresponds to what occurs, since the growth rate at the base of the apical dome is continued into the elongation of the stem below. From Green (1974).

edge of the meristem. This is consistent with minimal surface growth rates at the centre of the apex and with predominantly radial, or longitudinal, growth. However, precise rates of displacement would give a clearer picture since, even if there were a uniform growth rate over the whole surface of the apex, marks placed near the centre of the apex would be displaced radially at an accelerating rate simply because of the growth of the cells between the mark and the apical centre (Lyndon 1976).

In some apices there could be a region of much slower growth in the interior of the subapical part of the shoot meristem, analogous to the quiescent centre in the root. This would be found if the gradient of

Figure 3.6 The increasingly longitudinal polarity of growth down the sides of the apex of *Silene coeli-rosa* is visible as longitudinal files of cells on the flanks of the apex and in the developing primordia. Bar = 100μm.

growth rate from the summit to the flanks were not uniform, but increased rapidly on the flanks to a new constant value (Hejnowicz and Nakielski 1979) (Fig. 3.7). This would be consistent with the sort of structure seen in some gymnosperm apices in which there is a group of central mother cells that may be growing very slowly, as suggested by their thick cell walls (see Fig. 2.4).

It is easier to reconcile the cellular structure of the apical meristem and the growth rates within it if it is assumed that new cell walls form normal to one of the principal directions of growth. This enables the displacement lines of points within the apex to be described mathematically and possible growth rates in all directions to be calculated (Hejnowicz 1982). New cell walls do indeed usually seem to be normal to a principal growth direction. This led Lintilhac (1974) to propose the hypothesis that in dividing cells new cell walls form in the plane of least shear, which is usually perpendicular to the principal growth axis of the cells. The cell pattern, shape of the apical dome and distribution of growth in the apical dome appear to be consistent with maintaining minimal shear stress on the newly forming anticlinal and periclinal cell walls during growth (Hejnowicz 1982).

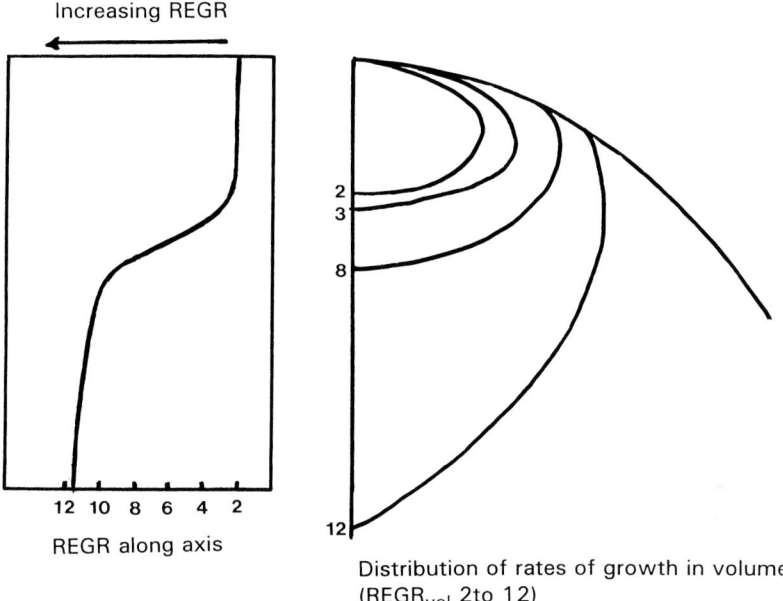

Increasing REGR

REGR along axis

12 10 8 6 4 2

Distribution of rates of growth in volume
(REGR$_{vol}$ 2 to 12)

Figure 3.7 If the growth rate down the apex is not uniform but increases suddenly, this would be consistent with the possibility of a slowly growing region just below the summit of the apex, as suggested by the structure of the *Pinus* apex (see Fig. 2.4). Data from Hejnowicz and Nakielski (1979).

3.3.1. *Why is there a tunica?*

These various models that reconcile apical dome shape with the gradient of apical growth rates and growth directions do not account for the existence of the tunica, which is essentially a skin of one or more cell layers in which there is no growth normal to the surface. All growth is in surface area and is therefore restricted to two dimensions instead of three, implying that the rate of growth of the cells is only two-thirds of what would be the case if the tunica structure did not exist and growth normal to the surface also occurred. The physiological or genetic basis for the existence of the tunica remains obscure. It is clearly not essential, since plants such as gymnosperms do not have it but, because it is such a widespread feature of the angiosperms, it presumably has some evolutionary advantage, perhaps in providing a mechanism to restrict growth rate, and hence the number of cell divisions, in the outermost cell layers.

3.3.2.   *What is the significance of the growth gradient?*

In almost all vegetative apices, whether having a single apical cell or a meristem, the cell or cells at the summit of the apex tend to divide more slowly than the cells on the flanks, or on the periphery of flatter apices, where the primordia are initiated. The main difference between mosses, gymnosperms and angiosperms is in the planes of division in the surface layers which result from the possession of a single apical cell or the presence, or not, of a tunica. The general occurrence of lowest growth rates at the summit and greatest on the flanks of the apex therefore raises the question: what is its significance?

A low rate of growth and cell division in the promeristem is characteristic of all meristems. The slowest dividing and growing cells are at the apical summit or centre in the shoot meristem (Table 3.4, Fig. 3.4), in the quiescent centre in the root meristem (Clowes 1976), and in the cambial initials in the cambium (Wilson 1964). In the shoot meristem the increasing growth rate down the flanks, coupled with the increasingly longitudinal polarity of growth, also allows for the elongation of the stem axis without any hiatus in growth, which would necessarily occur if there were tip growth, and growth rate was minimal or zero at the base of the apical dome (Fig. 3.5).

A maximal growth rate on the flanks or periphery of the dome may also be a factor facilitating primordium initiation in this region (see Chapter 6) and so also facilitating the modular, or metameric, growth of the shoot while preserving the terminal meristem. However, the apparent absence of a growth gradient in young pteridophyte apices (Table 3.6) and in some angiosperm reproductive apices (Table 3.5), but which also make primordia, implies that a lower rate of growth at the apical summit is irrelevant for formation of primordia. A lower growth rate at the summit may prevent primordia being formed there. When the rate of growth of the apical summit increases, and the differential in growth rates within the apex lessens or disappears, as in reproductive apices (Table 3.5), primordia can then form further up the apex and eventually on the summit itself, when carpels are formed.

It has been suggested that the existence of the slowest growing cells at the summit of the apex in the promeristem has evolved because it reduces the chance of errors in DNA replication in the progenitor cells (Hejnowicz and Nakielski 1979) and this seems the most likely explanation. This would also be equally valid for low rates of growth of other promeristem cells, of the root apex and the cambium.

3.3.3. *What determines the growth gradient and cell pattern?*

The principal growth axis is normally perpendicular to the main orientation of the wall microfibrils and the microtubules of the cell cortex (Green 1980). The microtubule and wall microfibril arrays normally form parallel to the new cell wall and so the cellular growth axis is normally perpendicular to the microfibril orientation (Green 1984) (also see Chapter 2). This set of mutually reinforcing hypotheses can explain the pattern of cells in the apex as a result of new cell walls being usually perpendicular to the local growth direction, and the longitudinal polarity of growth of the apical dome being a consequence of the distribution of growth rates within it. If the controlling factor of shoot apical dome shape and the cell patterns within it is the distribution of growth rates, then the key questions are: what determines the distribution of growth rates within the apical dome, and how does the growth rate gradient arise and how is it maintained? When cell cycle length differs in different parts of the apex is this by similar controls on the cell cycle in all plants? What are the features that sustain and generate initial cells? The attempt to answer these questions will be deferred to later chapters where it is linked to the questions of what controls the cell cycle (Chapter 4), what metabolic gradients exist within the apex (Chapter 5), and what controls growth of the apical surface when primordia are initiated (Chapter 6).

3.4. Control of apical growth

To what extent is the shoot apex autonomous or to what extent is its growth determined by the rest of the plant? The growth of the shoot apex can be affected by nutrition. If plants are starved of nutrients then the shoot apex becomes smaller; if the nutrient level is raised, then the shoot apex can become larger. In *Agropyron* plants given high nitrogen (nitrate) levels the size at the shoot apex was due to increased cell number rather than cell size (Rogan and Smith 1975).

In etiolated seedlings, the growth of the shoot apex slows down (Butler and Lane 1959) and in plants transferred to darkness for a time, it stops (Grose and Lyndon 1984). Whether this is simply because of a lack of assimilates from the rest of the plant or whether there is a direct light requirement of the apex itself for growth is not known.

3.4.1.   Control of growth of isolated apices

When shoot apices are excised and grown in culture, they have usually been found to require some growth substances in addition to sources of carbon, nitrogen and nutrients supplied by the medium. Shoot apical meristems, lacking leaf primordia, of the fern *Pteris* grew and produced leaves on a simple medium containing yeast extract (Michaux-Ferrière 1973). *Lupinus* apices grew best when given gibberellic acid and coconut milk (a source of cytokinins and sorbitol amongst other things) (Ball 1960b). A supply of auxin was essential for the growth of excised apices of *Nicotiana tabacum*, *N. glauca*, *Daucus carota*, *Tropaeolum majus* and *Coleus blumei* (Smith and Murashige 1970). Carnation (*Dianthus caryophyllus*) shoot apices required an auxin (NAA) but not gibberellin (Philips and Matthews 1964). More thoroughly investigated by Shabde and Murashige (1977), carnation apices were shown to be auxin- and cytokinin-requiring until they had a pair of expanding leaves together with two primordial leaf pairs, when plant growth-regulating substances were no longer required. A relatively high cytokinin–auxin ratio stimulated leaf initiation, whereas the reverse repressed it. In *Coleus*, IAA resulted in plant formation from an apex with only two pairs of primordia, and with IAA plus cytokinin one pair was enough. Cytokinin promoted initiation of rosettes of leaves (Smith and Murashige 1982). Apices of *Silene* with a single pair of leaf primordia grew and formed a flower without the addition of growth substances (Donnison and Francis 1993), but when auxin was added to flowering apices it promoted reversion to leafy growth (Donnison and Francis 1994). These data imply that auxin and cytokinin are required for leaf primordium initiation (see also Chapter 6). They also imply that maintenance of the growth rate of the apical dome may require auxin and cytokinin and that these can be supplied by the developing primordia and young leaves, which are known to be good sources of auxin (Goodwin 1978). Gibberellin does not seem to be essential for apical dome growth. But when supplied to *Taraxacum* apices it increased the length of the plastochron and proportion of juvenile leaves formed (Pomar et al. 1986).

What information we have, therefore, points to the shoot apex having no special requirements for growth in addition to the normal sources of carbon and nitrogen and mineral elements. The apical dome itself may require auxin and cytokinin, and these can be supplied by the young primordia and leaves.

3.5. Effect of apical growth rate on the rest of the plant

The growth rate of the shoot apex may, at least in some circumstances, determine the subsequent growth rate of the plant as a whole. The rate of growth in wheat up to collar initiation (as measured by rate of leaf appearance) was directly related to the rate of primordium initiation at the apex (Hay and Kemp 1990). The implication is that the effect of the environment in controlling the rate of development of the plant before flowering is primarily through its effect on controlling the rate of primordium initiation at the shoot apex. The shoot may also exert a control on the rest of the plant by means of correlative signals, such as growth substances and certain metabolites (Sachs 1991) but this is likely to be an effect of the larger shoot apical system, including the young leaves or flowers, rather than the apical meristem itself.

4

Cell Cycles

4.1. Growth fraction: the proportion of cycling cells

Growth fraction is a concept that has been used in describing the functioning of the root meristem. The idea is that not all the cells in the meristem are cycling and they can therefore be classified into cycling and non-cycling cells, the proportion of the total number of cells that are cycling being the growth fraction (Webster and MacLeod 1980). Cycling cells are recognized by their ability to incorporate radioactive label such as [^{3}H]thymidine so that their cycle time can be measured directly. If the mean cell doubling time measured by increase in total cell number is longer than the cell cycle measured by labelling, then this implies that there are unlabelled, non-cycling cells contributing to the longer, average cell doubling time, and the growth fraction is therefore < 100%.

In roots the concept of a growth fraction has obvious validity because of the presence of a group of cells, the quiescent centre, which are essentially non-cycling because they grow and divide very slowly relative to the rest of the meristem cells (Webster and MacLeod 1980). In some shoot apices there is also a clearly defined central zone of non-cycling or slowly cycling cells. In *Helianthus* and *Nicotiana* the cells of the central zone do not label with [³H]thymidine or, if the apex is excised so that cell division is stimulated throughout and [³H]thymidine is incorporated into all the cells, the central zone cells then become quiescent and they retain the label, because it does not become diluted through division and growth of the cells (Sussex and Rosenthal 1973, Langenauer, Davis and Webster 1974, Davis, Rennie and Steeves 1979). Even in a shoot apex in which there are mitoses throughout, there is the potential to accommodate non-cycling cells within it. Growing as it does in three dimensions, the shoot apex could contain non-cycling, and essentially non-growing, cells surrounded by growing and dividing cells. Such non-cycling cells would not be forced by geometrical considerations to enlarge and therefore would not be obviously distinct from their neighbours. Do apparently uniformly growing shoot apices contain intermixed populations of cycling and non-cycling cells?

In *Silene* shoot apices the growth fraction was assumed to be 100%, i.e. all the cells were cycling, because the mean cell doubling time (20–24h) in vegetative apices calculated from the rate of increase of cell numbers (Miller and Lyndon 1976, Lyndon 1977b) was almost the same as the cycle time (18–20h) found by radioactive labelling of a small fraction of the cells (Miller and Lyndon 1975, Francis and Lyndon 1978b). In other experiments, too, the mean cell doubling time of 18h was also almost the same as the cell cycle (16–17h) measured by labelling (Ormrod and Francis 1985). If the growth fraction had been significantly < 100% then mean cell doubling time should have been much longer. Similarly, in *Silene* apices exposed to one long day, which shortened the cell cycle to about 11h (Ormrod and Francis 1985), and in apices just before flowering, when the cell cycle shortened to about 10h, the mean cell doubling time measured by the rate of increase of cell numbers and the cell cycle time measured by labelling both gave similar values, indicating that the growth fraction was *c.*100% at all times (Miller and Lyndon 1975, 1976).

On the other hand, in *Sinapis* shoot apices the growth fraction was found to be much less than 100%, only about 32–41%. This would therefore represent the fast cycling cells, the other 59–68% being slow or non-cycling (Gonthier, Jacqmard and Bernier 1985). On transition to flowering, the growth fraction increased to 50–60% (Gonthier, Jacqmard and Bernier 1987). The labelled, cycling cells were scattered throughout the shoot meristem so that the apex was a mosaic of cycling and non-cycling cells. This suggested that the non-cycling cells were only temporarily out-of-cycle (Gonthier, Jacqmard and Bernier 1985).

However, the value obtained for the growth fraction depends very much on the methodology used to measure it. If there is a range of cell cycle times within the meristem then the size of the growth fraction will depend on the length of the experiment and the criteria used to classify a cell as cycling or non-cycling. What, then, in *Sinapis* is meant by a non-cycling cell? Would the methods used produce an arbitrary cut-off point? The first method used to distinguish between cycling and non-cycling cells was to label with [^{3}H]thymidine continuously for 72h. By then the proportions of labelled cells with > 3C and < 3C DNA amounts had stabilized, as would be expected if the meristem was asynchronous and all the cycling cells had progressed from the S-phase of the cycle to the next S-phase, i.e. all cycling cells would be expected to have become labelled. This method gave 33.5% for cycling and 66.5% for non-cycling cells (Gonthier, Jacqmard and Bernier 1985). A second method was to find the ratio between mean cell cycle length measured by pulse labelling (86h) (Gonthier, Jacqmard and Bernier 1985) and mean cell doubling time measured from the rate of accumulation of colchicine metaphases (206h) (Bodson 1975). This gave 41% for cycling and 59% for non-cycling cells. In the *Sinapis* apex non-cycling cells are therefore effectively defined as those cells which have not experienced the S-phase within 72h. Since S-phase in the cycling cells was 33h, non-cycling cells are those whose cell cycles must be > 100h. The non-cycling cells were scattered throughout the apex including the peripheral zone where primordia are initiated and where the cell cycle is fastest (Bodson 1975), so it was suggested that the non-cycling cells halt only temporarily and may revert to cycling before they leave the meristem (Gonthier, Jacqmard and Bernier 1985). An alternative explanation is that there is a range of values for the cell cycle and that only the fastest cycling cells have been recorded as cycling. However, the ratio of cells > 3C:<3C would not have been expected to stabilize as it did when only one-third of the cells were labelled, unless the faster cycling and slower cycling cells formed populations which were more or less discrete in behaviour even though not spatially distinct.

The cell cycle within the *Pisum* apex varies almost five-fold, from 15 to 69h, and there was no evidence of a division-free zone since divisions (though with different frequencies) were recorded throughout the whole apex (Lyndon 1973). However, the cells seem to fall into two populations (Fig. 4.1), those with cycles < 50h and those with cycles > 50h. If

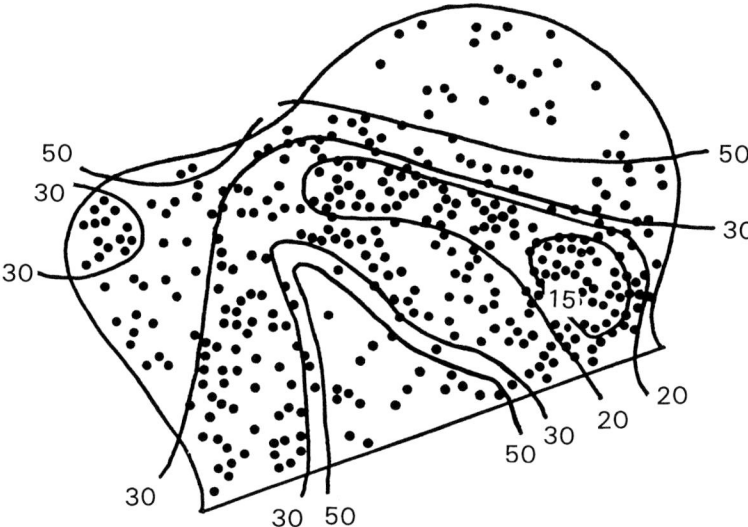

Figure 4.1 Contours of cell cycle times (h) in the pea (*Pisum*) apex (L.S.). The shortest cell cycles (15h) are in the region where the next primordium will emerge but also where procambial cells begin differentiating. From Lyndon (1973).

measurements of the growth fraction were made in *Pisum* by labelling for about 30h, approximately the mean cell doubling time (Lyndon 1973), one might expect to find that a substantial minority of the cells of the apex would be recorded as non-cycling. However, in the case of *Pisum*, these non-cycling cells would be expected to be localized in the central zone and the incipient pith, rather than scattered through-out the meristem as in *Sinapis*. There may, therefore, be a range of possibilities in the shoot apex. Some shoot apices (e.g. *Silene*) may consist entirely of cycling cells, others (e.g. *Pisum*, and *Helianthus* and *Nicotiana*) may have both fast and slow cycling cells but in different regions of the apex, and yet others (e.g. *Sinapis*) may consist not only of an intimate mixture of fast and slow cycling cells with a constant inter-change between the two populations, but also with some very slowly cycling cells at the apical summit, where the mean cell doubling time was 288h (Bodson 1975).

4.2. The cell cycle in shoot apices

In mosses, where each division of the apical cell produces a merophyte which gives rise to one leaf (see Chapter 1), the rate of division of the apical cell can be measured from the rate of leaf initiation, since the plastochron is equal to the doubling time of the apical cell, and this then equals the cell cycle length in the apical cell.

In higher plants other methods must be used. The cell cycle is not so easily measured in shoot apices as in root apices. The first problem is that the shoot apex, although appearing fragile, has a cuticle which hinders the absorption of solutions in contact with it. To label shoot apices with a solution of [^{3}H]thymidine, it has often been necessary to dissect away the enclosing young leaves to expose the apex and then to excise the youngest primordia. The label then enters through the wounded surfaces (Bernier and Bronchart 1963). This method has proved an effective way of labelling apices of *Sinapis* (Bernier and Bronchart 1963, Lyndon, Jacqmard and Bernier 1983), *Papaver* and *Perilla* (Bernier and Bronchart 1963), *Pisum* (Lyndon 1973), and *Silene* (Miller and Lyndon 1975, Francis and Lyndon 1978b). The obvious question is whether this drastic treatment affects the growth of the apex and alters the cell cycle. Treated apices can grow on normally (Miller and Lyndon 1975) and no obvious effects on mean cell generation times or mitotic indices have been noted. Less drastic removal of tissues near the apex has proven equally effective in *Sinapis*, where only the 5mm leaf needed to be removed (Gonthier, Jacqmard and Bernier 1985) and in *Lolium*, where [^{3}H]thymidine was injected into the leaf sheath cavity just above the apex (Ormrod and Bernier 1990) or a window was made through the leaf sheaths to expose the apex (Houssa et al. 1994).

The second problem is that the relatively long cell cycles and low mitotic indices (characteristically only a few percent) make pulse labelling experiments difficult. When the mitotic index is only about 2% and the whole of each shoot apex consists of only about 1000 cells, scoring changes in the percentage of labelled mitoses after a pulse of [^{3}H]thymidine calls for many samples. The number of samples and replicates possible is dictated by the number of apices it is possible to prepare (often by dissection and removal of young primordia under the microscope) to receive the labelled solution. Despite these problems, the percent labelled mitoses (PLM) method has been used successfully to measure the cell cycle and its phases in *Silene* (Ormrod and Francis 1986a, Taylor and Francis 1989) and *Sinapis* (Gonthier, Jacqmard and Bernier 1985).

To make maximum use of the available experimental material a double labelling method was devised (Miller and Lyndon 1975). It was based on that of Wimber and Quastler (1963) and is particularly useful for shoot apices because every labelled nucleus contributes to the data, and not just labelled mitoses. This method, which requires a minimal number of samples, and is the method of choice if the number of apices available is small, has been used successfully for *Silene* (Miller and Lyndon 1975, Francis and Lyndon 1978b) and for *Chrysanthemum* (Nougarède and Rembur 1977).

4.2.1. Cell cycle phases: control points

The phases of the cell cycle can be measured by pulse labelling (PLM), double labelling or calculated from the proportions of cells in different cell cycle phases if the proportion of dividing cells is known or can be assumed. Such measurements have been made in the different regions of the apex in five species (Fig. 4.2). The shortening of the cell cycle from the central zone of the shoot apex to the peripheral zone seems to be primarily because of a reduction in the length of the G_1-phase. However, G_2-phase was also greatly reduced in *Pisum* and *Isoetes* and to a smaller degree in *Rudbeckia*. Also, S-phase was reduced in *Rudbeckia*, *Pisum* and *Helianthus*. In *Helianthus* and *Chrysanthemum* the major reduction was in G_1, mainly perhaps because it tends to be the longest phase anyway. In *Pisum* and *Rudbeckia* (and possibly also *Isoetes*), all phases of the cycle (except for M) reduce, as though the whole cycle were elastic.

Once they begin to be displaced down from the summit or away from the centre of the apex, cells will tend to find themselves in different regions of the apex in successive plastochrons and eventually will finish up in the primordia or stem. If the plastochron is of similar length to the cell cycle, then in the course of a single cycle cells will be displaced into regions of the meristem where the cell cycle is of different length, usually shorter (Lyndon 1973). Particularly when cells are being displaced down the flanks of the apex, the cell cycle will tend to be continually changing. Such continual adjustments to the length of a cell's cycle would be facilitated by having several control points rather than just one, and this seems to be the case for *Pisum* and *Rudbeckia* in which all the cell cycle phases except mitosis are variable in length.

The way in which the length of the cell cycle alters differs in detail in different plants but the main control point seems to be at the entry to S-phase, since the greatest difference between the cell cycles in the central and peripheral zones of the apex is in the length of G_1 (Fig. 4.2). The

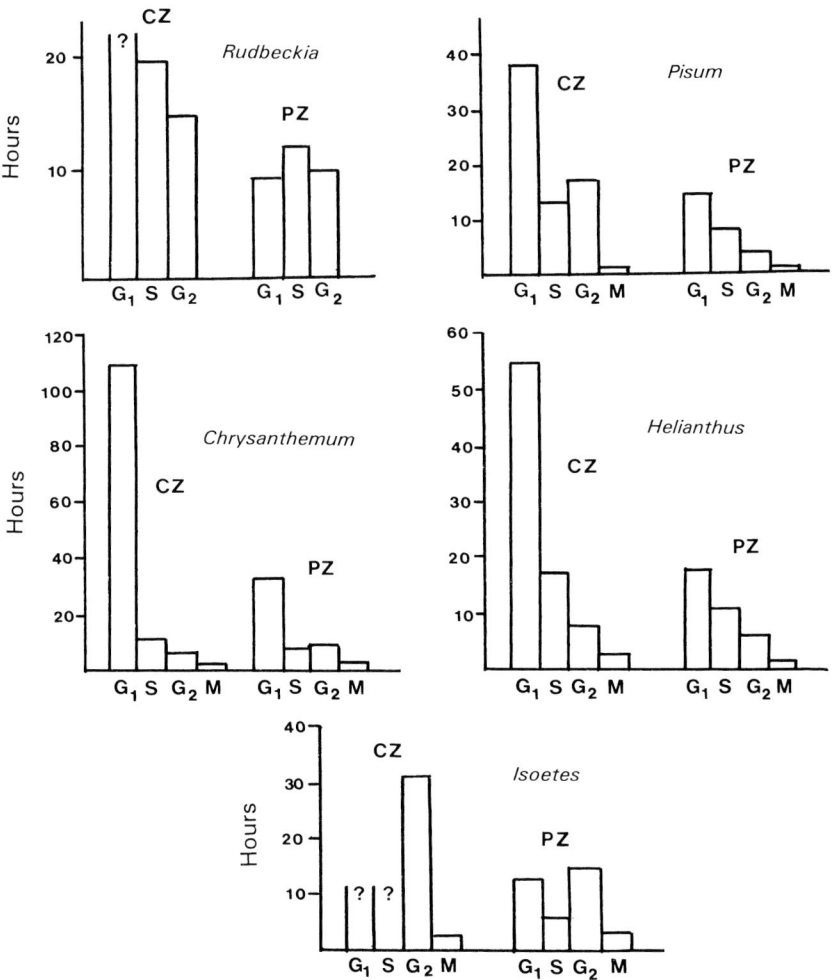

Figure 4.2 Lengths of the phases of the cell cycle in the central zone (CZ) and peripheral zone (PZ) (flanks of the apex) in four angiosperms, and a pteridophyte (*Isoetes*). Data from Jacqmard (1970) (*Rudbeckia*); Lyndon (1973) (*Pisum*); Nougarède and Rembur (1978) (*Chrysanthemum*); Marc and Palmer (1984) (*Helianthus*); and Michaux (1969) (*Isoetes*).

shortening of S on the flanks of the apex (Fig. 4.2, PZ) shows that the rate of DNA synthesis also tends to increase. This could result from either increased rates of DNA synthesis by an increase in the rate of DNA replication fork movement or by an increase in the numbers of replication origins (Ormrod and Francis 1986b), but we have no data for the different regions of the shoot apex. However, the application of

Table 4.1. *Cell cycle (CC) phases in* Polytrichum *(Hallet 1977).*

Cell type		G_1	S	G_2	M	G_2/G_1	CC (h)
		Percentages of cells in:					
Vegetative	Apical cell	38	8	53	1	1.46	360
	Inactive sec. initial	94	–	6	–	0.06	–
Reproductive	Apical cell	39	7	51	3	1.31	144
	Active sec. initial	37	12	51	n.a.	1.38	n.a.

		G_1	S	G_2	M	CC
		Hours				
Vegetative	Apical cell	136	29	191	4	360
Reproductive	Apical cell	56	10	74	4	144

n.a., not available.

cytokinin (benzyladenine) to shoot apices increases the rate of DNA replication by doubling the number of replication origins in the DNA in *Sinapis* (Houssa, Jacqmard and Bernier 1990) and also by increasing the rate of replication fork movement in *Lolium* and *Lycopersicon* (Houssa et al. 1994). Whether this mimics what happens naturally as cells are displaced down the flanks of the vegetative shoot apex is not yet known.

The only data for ferns suggests that the main control point may be at the entry to M since the cell cycle in the apical cell of *Pteris* is characterized by a long G_2 (60% of the cycle) (Michaux 1971). In the moss *Polytrichum*, the proportions of apical cells in the various phases of the cell cycle are similar for active cells, the greatest proportion being in G_2 (Table 4.1), but inactive initials are nearly all arrested in G_1, as are the apical cells of inhibited lateral branches, which nearly all have the 1C DNA amount (Hallet 1974). Because these values refer to single apical cells and not populations within a meristem, the proportions of cells in each phase are a direct function of the time spent in each phase. G_2 is therefore the longest phase in active initials, and because the proportions are similar for active cells with different cycle lengths this implies that all phases of the cell cycle (except mitosis) are shortened when the cell cycle shortens, as it does on the transition from vegetative to reproductive (Table 4.1). This implies that the control points at the entries to S and M are equally important in the moss. However, data are not available to

Table 4.2. *Changes in the mean cell cycle time (CC) with meristem age in vegetative plants.*

Plant	CC (h)	References
Mosses		
Polytrichum (juvenile, 30 day)	34	Hallet (1977)
Polytrichum (adult)	360	Hallet (1977)
Leucobryum (juvenile, 30 day)	35	Hallet (1977)
Leucobryum (adult)	136	Hallet (1977)
Pteridophytes		
Polypodium (young axial zone)	62	Michaux-Ferrière (1981a)
Polypodium (adult axial zone)	144	Michaux-Ferrière (1981a)
Polypodium (young lateral zone)	61	Michaux-Ferrière (1981a)
Polypodium (adult lateral zone)	78	Michaux-Ferrière (1981a)
Isoetes (young axial zone)	21	Michaux-Ferrière (1980)
Isoetes (adult axial zone)	> 53	Michaux (1969)
Isoetes (young lateral zone)	20	Michaux-Ferrière (1980)
Isoetes (adult lateral zone)	36	Michaux (1969)
Angiosperms		
Lupinus (7 days old, plastochron 10)	21	(calculated from):
Lupinus (14 days old, plastochron 14)	24	Sunderland and Brown (1976)
Lupinus (21 days old, plastochron 18)	33	Sunderland and Brown (1976)
Agropyron (vegetative 2 days old)	66	Smith and Rogan (1979)
Agropyron (vegetative 25 days old)	74	Smith and Rogan (1979)
Agropyron (vegetative 80 days old)	624	Smith and Rogan (1979)

show whether this is also the case for faster dividing cells adjacent to the apical initial.

Mitosis itself is similar in length (typically 1–3h; see Table 3.3) in mosses and in higher plants, and is the one phase of the cell cycle that tends to remain constant except when temperature changes, when all phases of the cycle are affected (Francis and Barlow 1988). But in the mosses, with long cell cycles in adult plants (Tables 4.1, 4.2), this means that a very low mitotic index (of the order of 1–2%) is what would be expected in cycling cells and does not indicate non-division or inactivity of the apical cells.

4.2.2. Possible mechanisms of control of the cell cycle in vegetative apices

The molecular controls of the plant cell cycle, involving cell division cycle (*cdc*) genes and protein kinases, appear to be the same as those common to other organisms, and details can be found elsewhere (Francis 1992, Francis and Halford 1995). In plants, the cell cycle can also be regulated, probably indirectly, by plant growth substances.

Cytokinin applied to shoot apices stimulates cell division as shown by an increase in the mitotic index or the labelling index (Jacqmard, Houssa and Bernier 1994). The natural cytokinin, zeatin, gave a transient response in the *Sinapis* shoot apex, whereas benzyladenine, which presumably persisted longer in the tissues, gave successive peaks of mitotic index about 20h apart (Bernier et al. 1977). It is not clear whether this was because the cell cycle was shortened to 20h or whether the cytokinin had simply induced successive waves of mitosis by shortening G_2. Cytokinin also shortens the S-phase of the cycle by activation of latent DNA origins so that replicon size is halved and the overall rate of DNA synthesis is therefore doubled (Houssa, Jacqmard and Bernier 1990, Houssa et al. 1994).

Abscisic acid (ABA) had the opposite effect to cytokinins in *Sinapis* apex, by inactivating some of the DNA origins so that mean replicon length increased from 7.5 to 32.5μm, although the rate of replication (rate of fork movement) was unchanged (Jacqmard, Houssa and Bernier 1995).

Gibberellins can also affect the cell cycle in shoot apices. In *Perilla*, application of gibberellic acid to the shoot apex increased mitotic index in all regions of the meristem. In *Rudbeckia* gibberellic acid increased mitotic index six-fold in the central zone and three-fold in the pith but had no effect on the mitotic index of the peripheral zone and pith-rib meristem (Bernier, Bronchart and Jacqmard 1964). DNA synthesis was stimulated before mitotic index so that gibberellin appears to act in *Rudbeckia* by stimulating entry into both S and M (Jacqmard 1968). Similarly in *Silene armeria*, gibberellic acid transiently increased the mitotic index and proportion of cells in G_2, with a corresponding reduction in the proportion of cells in G_1, indicating a stimulation of entry into S and M (Besnard-Wibaut, Noin and Zeevaart 1983).

Whether concentrations of cytokinins and gibberellins, or sensitivity to them, actually change in the shoot apex, and whether this might control the cell cycle there, is not known. Sugar concentration is not a likely controlling factor since the concentrations of sucrose were similar throughout the *Sinapis* meristem (Bodson and Outlaw 1985). There is no other evidence for gradients of metabolites, apart from the gradient of RNA and ribosomes indicated by the zonation patterns of apices (see Chapter 5).

The growth gradient and differences in the cell cycle throughout the apex could be maintained by either a chemical or a physical gradient. A chemical gradient could perhaps prevent growth at the summit and/or promote it on the flanks of the apex. It would be simpler to postulate an inhibitor that decreases in concentration and effectiveness as it diffuses away from its main production site at the summit. Growth could perhaps be limited by mechanical constraint if the apical summit surface is stiffer and less deformable than on the flanks (or has the wrong surface micro-structure configuration). This would then pose the question of how cellular growth rates and the control of the cell cycle might be exercised by physical constraints on the growth of the cells. The problem becomes one, not of the growth gradient, which is then secondary, but of the deformability of the surface and its ability or otherwise to form primordia. Even when there is no growth gradient, as in young pteridophyte seedlings, leaves are still not formed at the summit of the meristem. But these pteridophyte sporophyte meristems form leaves that are often sporophylls, bearing sporangia, and so may be regarded as comparable with the reproductive apices of higher plants in which the growth gradient may have disappeared, but which also do not form primordia at the apical summit except as their last act.

There are undoubtedly internal controls on the growth rate and the cell cycle in shoot apical meristems, as a number of examples suggests. The slow growth of the oil palm and plants of the Brazilian cerrado, tropical plants growing in apparently favourable environments, has already been mentioned (see Section 3.1.3). The slow growth rate of alpine plants has also been suggested as being due to internal controls on cellular growth rate rather than due to the environment directly (Körner and Menendez-Riedl 1989). The nature of the internal controls on shoot apical growth has yet to be discovered.

4.2.3. *Control of organelle replication*

To know about the rates of organelle replication in shoot apices, information is needed about changes in the numbers of organelles. The only definitive data available is for the *Pisum* shoot apex (Lyndon and Robertson 1976). As cells are displaced from the central zone to I_2 and then to I_1 and into the leaf primordia, the numbers of mitochondria, chloroplasts, dictyosomes and microbodies per cell all increased by about 50%. The fact that they increased at all showed that their rates of replication outstripped the rates of cell replication. In the cells that formed the pith–rib meristem the number of chloroplasts per cell, 10 or 11, remained constant, but the number of mitochondria increased from about 58 to 83, dictyosomes from 15 to 20, and microbodies from 4 to 12. The replication of different types of organelles, and of the cells, was therefore controlled differentially. In the cells that would form the incipient axillary bud, the numbers of organelles per cell did not increase, implying that here organelle replication kept in step with the cell cycle (Lyndon and Robertson 1976). How the rates of organelle replication are controlled in relation to the cell cycle is not yet known.

4.2.4. Ageing of the meristem

As vegetative apices age, the cell cycle tends to lengthen (Table 4.2) and growth rate therefore decreases. In mosses the cell cycle of the apical cell increases with age. Since it is calculated from the phyllochron, this implies that the cell cycle throughout the whole shoot meristem lengthens as the leafy gametophyte ages, although it may shorten again when the plant becomes reproductive, as in the apex of *Polytrichum* and in the young archegonia (Hallet 1974), in the same way as it does in the meristem of higher plants on floral induction.

In the pteridophytes *Polypodium* and *Isoetes*, the cells of the flanks continue growing at a rate similar to that in the young apex whereas the cells of the apical summit slow down (Table 4.2). It is not known whether this also happens in higher plants.

In *Agropyron* cell doubling times lengthened considerably, from 67h in young seedlings to > 600h in 86-day-old plants (Smith and Rogan 1979). Although the cell cycle may lengthen with ageing in a vegetative meristem, especially as the winter approaches, as in conifers the cell cycle will resume again the next year and shorten as summer approaches (Cannell 1976). In herbaceous plants the cell cycle shortens again later in the same season at the onset of flowering (see Section 4.4).

4.3.   Environmental effects on the cell cycle in vegetative plants

Within the limits for growth of a plant, an increase of temperature shortens the cell cycle (Francis and Barlow 1988). Most measurements have been done on roots but the shoot meristem probably behaves in the same way. In *Silene* plants grown at 13°C the cell cycle in the vegetative apical dome was 57h, twice as long as that at 20°C (26h) (Lyndon 1979a). At 27°C the plants were at the upper limit of their temperature tolerance and the cell cycle was much longer (93h) than at the lower temperatures. When the transition to flowering had taken place the cell cycles in the flower were shorter, but still longer at 13 than at 20°C, but at 27°C the cell cycles in the flower were now comparable with those at 13°C. In *Dactylis*, the cell doubling time in the vegetative apex of a Portuguese population was reduced by about 40% by raising the temperature from 10 to 30°C. However, temperature had little effect on apex cell doubling time in a Swedish population, nor in either population subjected to elevated CO_2 concentrations (Kinsman et al. 1996).

Light quantity and quality can also affect the cell cycle in the shoot apex. When *Sinapis* plants were chilled to 2–3°C, mitotic activity ceased and resumed again when the plants were returned to 20°C; it resumed sooner if the plants were then in the light (Bodson 1970). However, high irradiance given to vegetative plants did not appreciably affect the cell cycle in *Sinapis*, although it increased sucrose concentrations in the shoot meristem (Bernier et al. 1993, Havelange and Bernier 1983). Shoot apices of *Chrysanthemum* plants grown in a photon flux density of 200μmol m^{-2} s^{-1} had mean cell cycles in all regions of the apex which were shorter than those of plants grown in 70μmol m^{-2} s^{-1} (see Table 3.4) (Nougarède et al. 1990), and the G_2/G_1 ratios were higher at the higher irradiance showing that the cell cycle was shorter mainly because of a shorter G_1. Whether the shorter cell cycle in *Chrysanthemum* was related to an increased carbohydrate supply to the apex or was the result of some more direct effect of light remains an open question.

Light quality, specifically red (R) and far-red (FR) light, can also affect the cell cycle. Rice seedlings were grown in darkness for 3.5 days, then exposed to R or FR light for 16 or 40h, and the mean cycle time in the meristem was measured from the accumulation of colchicine metaphases (Rolinson and Vince-Prue 1976). After 16h, the cell cycle in most parts of the shoot apex was shortened by R light but lengthened by

FR light. However after 40h, FR light had also shortened the cell cycle in most parts of the apex.

R and FR light have also been shown to have distinctive effects on the cell cycle at the transition to flowering, and these effects will be considered below (see Section 4.4.2).

4.3.1. Diurnal rhythms

Diurnal rhythms of mitotic index have been found in a number of plants and data have been summarized and tabulated by Edgar (1961) and Lyndon (1976). Although they do not give direct information about the cell cycle, changes in mitotic index imply changes in the degree of synchrony of cell division and this means that the cell cycle has been affected, probably shortened, albeit perhaps temporarily and without there necessarily being major alterations to the length of the cell cycle. Indeed, even lack of clear changes in mitotic index does not rule out changes to the cell cycle (Harte and Lindenmayer 1983). Except for *Tradescantia* and *Trifolium* shoot apices (Denne 1966b, c), and *Lonicera* (Edgar 1961) in which there were two peaks of mitotic index during the day (07:00 and 11:00 hours; the maximum was 19.0% at 07:00 hours) in addition to two at night (23:00 and 01:00 hours), all peaks of mitotic index have been found at night. This could be because of the interaction with carbohydrate availability, cell division being stimulated if carbohydrate concentration becomes lower during darkness, as suggested by Lyndon (1976) and as has been shown for the barley apex (Cottrell and Dale 1986). Alternatively it could be an interaction with water status. If the plants are stressed during the light period, division may become reduced but pick up again (indeed peak) when turgidity is restored during the dark period. There is, however, little or no firm information on the effect of water stress on cell division itself, although cell enlargement is very sensitive. In the absence of exact data about water status and light energy it is difficult to know precisely how these environmental variables affect mitotic index. Edgar's (1961) *Lonicera* hedge was, of course, growing outside and samples over 24h were taken over a sunny day in summer when the temperature rose from a chilly 2.2°C (at 04:00 hours) to 23.5°C at 13:00 hours. There is no mention of rain so presumably it was a dry day and it seems more than likely that the plants built up a water deficit during the day. There are simply too many unknowns in the data for diurnal rhythms to be sure what factors may be reponsible for the mitotic index fluctuations and so for affecting the cell cycle. Any diurnal rhythms may also be perturbed when the cell cycle is altered at the transition to flowering (see below), although in *Rudbeckia* the phase of the diurnal rhythm of mitotic index was essentially unchanged when flowering was induced by long days, but the mitotic index was higher in the induced plants (Milyaeva and Chailakhyan 1981).

4.3.2. Dormancy

It seems a general rule that the cell cycle in dormant shoot apices is arrested in G_1 (see Tepfer, Nougarède and Rondet 1981 for references). In the winter, the dormant shoot apex of the ash (*Fraxinus*) shows no mitoses and does not synthesize DNA (Cottignies 1974). The nucleolus shrinks, the granular zone of pre-ribosomes disappears and it also no longer synthesizes RNA (Cottignies 1977). All the cells are then in G_1, so that all cell cycles are stopped at or before the entry to S (Cottignies 1983). Also, in the inhibited axillary bud at the sixth youngest node in *Pisum*, most of the nuclei are 2C and therefore arrested in G_1 (Nougarède and Rondet 1975).

At the breaking of dormancy, in the Spring, the cells in the whole *Fraxinus* apex resume cycling in a weakly synchronous manner, as would be expected from cells all at the same point of the cycle, and then seem to become asynchronous as time goes on (Cottignies 1979). In the shoot apex of *Xanthium* embryos coming out of dormancy, after 28h of rehydration of the seeds, nuclei throughout the apex are all in G_1. The first mitoses become visible 46h after rehydration, 9–12h after the first 4C nuclei begin to be seen, indicating $S + G_2$ to be 9–12h (Rembur 1970).

4.4. Transition to reproductive growth

At the transition to flowering or reproductive growth, the rate of growth of the shoot apex increases, and the cell cycle therefore shortens, even if only transiently (Table 4.3, and see Table 3.5). In most plants the apical dome has been treated as a single entity for measurements of cell cycle length, but in *Datura* and *Sinapis* it has been shown that the cell cycle shortens proportionately in both the central zone and on the flanks so

Table 4.3. *Mean cell cycle times in vegetative and early floral shoot apices.*

Plant	Mean cell cycle (h)		Reference
	Vegetative	Floral	
Lupinus	48	34	Sunderland (1961)
Secale	50	31	Sunderland (1961)
Datura	36	26	Corson (1969)
Sinapis	157	25	Bodson (1975)
Silene	20	10	Miller and Lyndon (1975)
Triticum	41	22	Griffiths (1981)
Ranunculus	56	47	Meicenheimer (1979)
Epilobium	45	45	Meicenheimer (1982)
Impatiens	24	8	Battey and Lyndon (1984)

Table 4.4. *Cell cycle and its phases (h) in central and peripheral zones of* Helianthus annuus *in young (8 day) vegetative apices and in apices (16 day) beginning flower formation (Marc and Palmer 1984).*

Phase	Central zone		Peripheral zone	
	Veg (8 day)	Floral (16 day)	Veg (8 day)	Floral (16 day)
G_1	54.7	20.8	17.8	23.6
S	17.4	6.5	11.1	7.2
G_2	7.7	4.1	6.1	5.7
M	2.2	1.8	2.0	2.2
CC	82.5	33.2	37.1	38.7

that although the cell cycle is shortened throughout the apex the growth gradient is maintained in the floral apex (Corson 1969, Bodson 1975). The cell cycle at the summit of the vegetative apex is typically twice as long or longer than on flanks but may reduce to approximate equality on flowering (see Table 3.5). In *Helianthus*, the only plant for which values for cell cycle phases have been obtained for the different regions of the apical meristem in transition to flowering (Marc and Palmer 1984), the reduction in the length of the cell cycle takes place only in the central zone, and involves 50–60% reductions in G_1, S and G_2 so that the cell cycle becomes essentially uniform throughout the apex (Table 4.4). Indeed, in all five angiosperm species in which the cell cycle phases have been measured, when the cell cycle shortens on flowering, G_1, S and G_2 are all reduced, the greatest proportional reduction usually

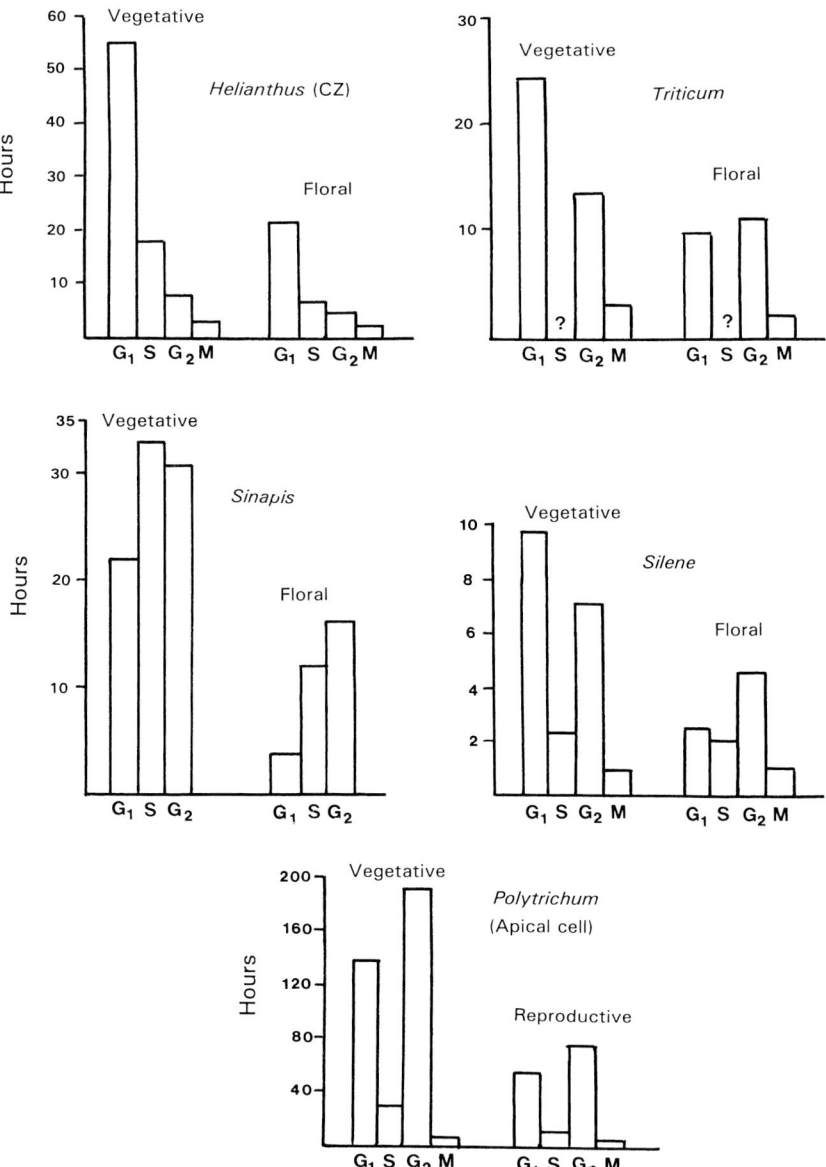

Figure 4.3 Lengths of the phases of the cell cycle in vegetative and flowering apices of four angiosperms, and the apical cell of vegetative and reproductive plants of *Polytrichum* (a moss). Data from Marc and Palmer (1984) (*Helianthus*); Griffiths (1981) (*Triticum*); Gonthier, Jacqmard and Bernier (1987) (*Sinapis*); Francis and Lyndon (1979) (*Silene*); and Hallet (1977) (*Polytrichum*).

being in G_1 (Fig. 4.3). This is also true of the moss *Polytrichum* on transition to reproductive growth (Table 4.1). In *Silene* apices, although S seems unchanged at the end of induction (day 8/9) and just as the sepals are being initiated (Fig. 4.3), this follows a marked shortening of S on the first day of floral induction (Fig. 4.6). On transition to reproductive growth or flowering, the only phase of the cell cycle remaining unaltered is M, although even it may also be shortened in *Helianthus* and *Triticum* (Fig. 4.3). It therefore seems characteristic of the switch to flowering that there is not one but several control points operative in the cell cycle.

A new method of investigating the relationship of the cell cycle to apical growth and development is now available with the use of transgenic plants. The yeast mitotic inducer gene, *cdc25*, has been transferred to tobacco and resulted in smaller cells and premature flowering (Bell et al. 1993). It will be fascinating to know what effects this gene has on the cell cycle and on the structure and functioning of the shoot apex, especially at the transition to flowering.

4.4.1. *Synchrony*

At the same time as the cell cycle becomes shorter on the transition to flowering, cell division also becomes synchronized. This was first noticed in *Sinapis*, in which an inductive photoperiod caused two waves of mitosis before the beginning of flower morphogenesis (Fig. 4.4) (Bernier, Kinet and Bronchart 1967). The mitotic peaks were 36h apart and this corresponded to the new, shorter cell cycle length in the induced apices (Bodson 1975). The proportion of cells with the 4C amount of DNA increased just before the first mitotic peak showing that most of the cells had moved into G_2 (Jacqmard and Miksche 1971). After the first mitotic peak most cells were now in G_1 (Jacqmard and Miksche 1971), so the G_2/G_1 ratio reached a low, then there was a peak of DNA synthesis, followed by a peak in the G_2/G_1 ratio, because most cells were now again in G_2, and then the next peak of mitosis (Fig. 4.4). The increase in the rate of DNA synthesis in *Sinapis* was because of an increase in the number of DNA initiation points so that replicon number was doubled but with no change in the rate of DNA initiation fork movement (Jacqmard and Houssa 1988). Changes in enzyme activities and in the numbers of organelles were also in step with what would be expected of cell cycles that had become synchronized (Lyndon and Francis 1984).

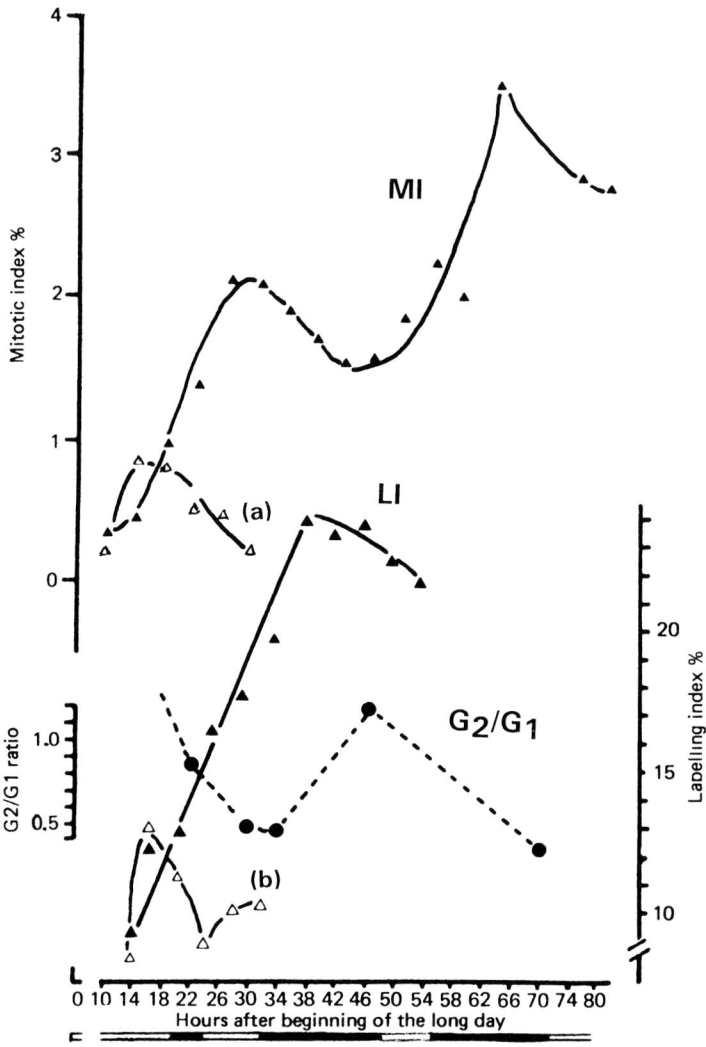

Figure 4.4 Broadly synchronous cell cycle on induction of flowering in *Sinapis*. The first peak of mitosis (MI) is followed by a minimum in the G_2/G_1 ratio (i.e. most cells now in G_1), then a peak in labelling index (LI), (% of cells incorporating [^{3}H]thymidine, indicating DNA synthesis), then a maximum in the G_2/G_1 ratio (i.e. high proportion of cells in G_2) with a minimum in mitotic index, followed by the second mitotic peak at about 64h. (a) Mitotic index; and (b) labelling index of control plants kept in short days and not induced to flower. Data from Bernier, Kinet and Bronchart (1967) and Jacqmard and Miksche (1971).

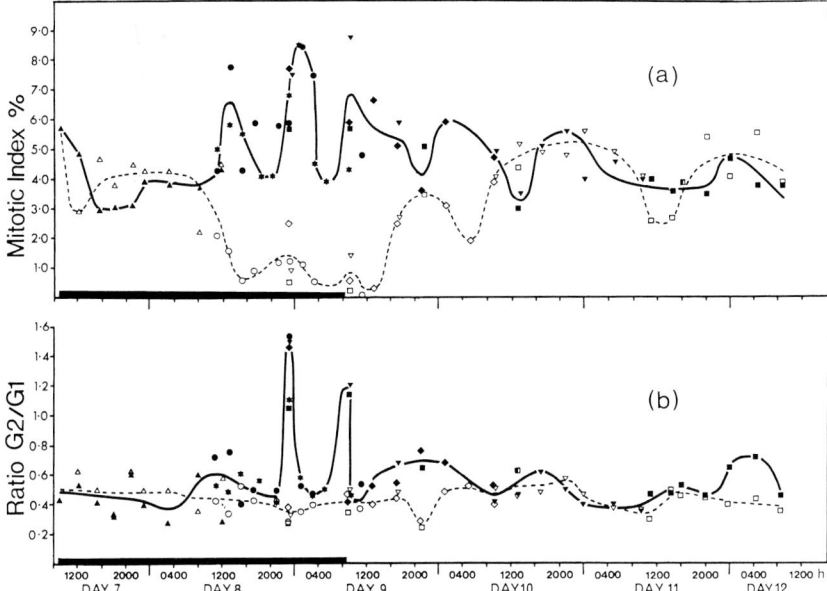

Figure 4.5 In *Silene* synchronous cell cycles, as shown by the peaks in mitotic index (a, solid line), normally occur before sepal initiation at the beginning of day 10 (after the beginning of long-day induction). This is confirmed by the peaks in the G_2/G_1 ratio (b, solid line) which just precede the mitotic peaks. When the plants are placed in 48h continuous darkness (black bar) after long-day induction, synchronous division is eliminated, as shown by the reduction of mitotic index (a, dashed line), as growth and division are inhibited, and the lack of G_2/G_1 peaks (b, dashed line), yet flowering is unaffected, though delayed by 48h, sepals not being initiated until the beginning of day 12. From Grose and Lyndon (1984).

Synchronization has also been demonstrated in *Silene*, at the end of floral induction and just before the initiation of the sepals (Francis and Lyndon 1979). Synchronization does not begin until day 8 after the beginning of floral induction (Fig. 4.5) (Grose and Lyndon 1984), although the cell cycle had already shortened as shown by the rapid increase in cell number which starts on day 7 (Miller and Lyndon 1976, Lyndon 1977b). The cycle shortened by the reduction of G_1 and G_2, but especially G_1 (Fig 4.3). As shown by the peaks in mitotic index and in the G_2/G_1 ratio (Fig. 4.5), strong synchrony persists for only two cell cycles and then dissipates.

Evidence for synchronization has also been found in *Xanthium*, in which a long night induced successive peaks of mitosis accompanied by the expected shifts in the proportions of cells in G_1 and G_2 (Jacqmard et al. 1976). Synchronization is probably a regular feature of the transition

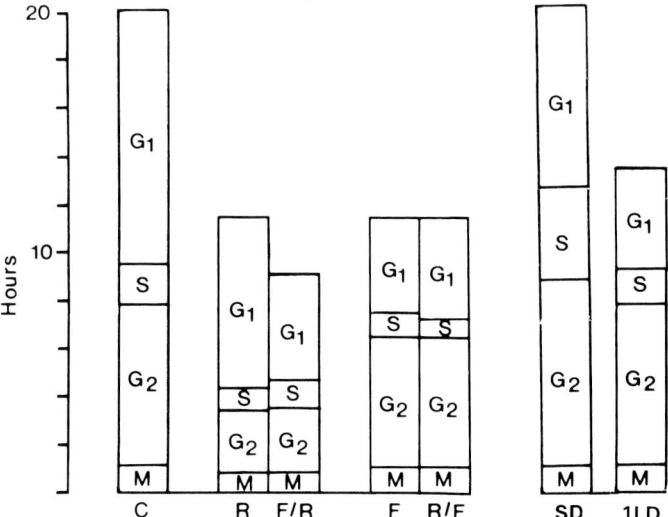

Figure 4.6 Effect of red (R) and far-red (F) light on the cell cycle in *Silene*. Although both R and F light shorten the cell cycle, compared with the untreated controls in short days (C, SD), only F light mimics the effect of 1 long-day (1LD; white light), shortening G_1 and S. With R light, G_2 is also shortened. Data from Francis (1981b) and Francis and Lyndon (1978b).

to flowering, as indicated by the peaks of mitotic index at this time which have also been noted in *Pharbitis*, *Chenopodium* and *Lolium* (Bernier 1971), but its significance to flowering is not clear (see Chapter 9).

4.4.2. Effects of R and FR light

It was discovered that in *Silene* the rate of growth of the apex could be stimulated by only one long day (Miller and Lyndon 1976, Francis and Lyndon 1978a), that the cell cycle was shortened from 20 to 13h (Francis and Lyndon 1978b), and that the number of cells in G_2 had already increased only 1h after the beginning of the change in light regime (Francis 1981a). When plants were treated to alternate long and short days, the G_2 increase occurred only at the beginning of the long day extension and not in the plants in short days (Francis 1981a). These observations led to the effects of light on the cell cycle in *Silene* being examined in more detail.

The reduction of the cell cycle from 20 to 13h by a long day could be mimicked by 5min of R or FR light (Fig. 4.6) (Francis 1981b, Ormrod and Francis 1985). The main changes are a reduction of G_1 and S. The

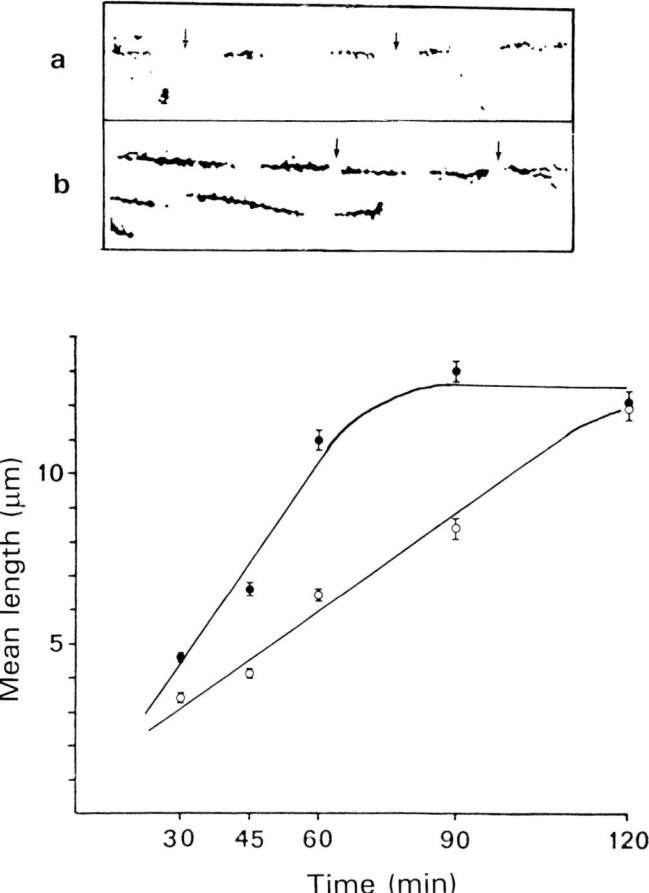

Figure 4.7 DNA replication in shoot apices of *Silene* plants exposed to 1-long day (1 LD), which shortens the cell cycle. The rate of replication of DNA molecules, measured as the rate of replication fork movement, is greater in 1 LD (a) than in short-days (SD) (b) as shown by the greater lengths of DNA that have become labelled in (a) than in (b) after 60 min; and by the greater rate of fork movement in 1 LD (closed circles) than in SD (open circles). Data from Ormrod and Francis (1986b).

reduction of S is by an increase in the rate of DNA replication fork movement (Fig. 4.7), which looks as if it is triggered immediately by the increase in FR/R ratio (Ormrod and Francis 1986b). There is no change in the number of replicating origins in the DNA and therefore no change in replicon length. The reduction in G_1 results in an increase in the G_2/G_1 ratio, which is caused by FR but not by R light, and the effect of FR light is reversed by R light, confirming a phytochrome response

(Ormrod and Francis 1986a). The timing of the FR light exposure in shortening the cell cycle is crucial: if, at the end of the period of high photon flux, the 5min of FR light is substituted by 20min of darkness, the change in the cell cycle is prevented. Subsequent exposure to light does not reverse the effects of this short period of darkness (Ormrod and Francis 1987). The relationship of changes in the cell cycle to flowering are discussed further below (Chapter 9).

4.5. The cell cycle and growth rate

The precise way the cell cycle is controlled clearly differs according to the position of the cells in the meristem, the stage of development of the meristem and the particular experimental conditions (e.g. R/FR light) to which the plants are exposed. The question is whether the cell cycle controls growth rate, or growth rate controls the cell cycle. Synchronization perhaps answers this question: if the cell cycle changes to cause synchronization but the growth rate is unaltered, as seems to happen in *Silene*, then the implication is that the cell cycle can change without altering growth rate. In the shoot apex, where cells are displaced down the dome and the cell cycle may be changing continually, it is hard to think of the cell cycle as the factor controlling growth rate. It seems much likelier that the cell cycle is reponding to the growth rate environments through which the cells pass. If so, then cell division is, in the short-term, an enabling factor rather than a cause of growth. Only in the longer term, when cell size is maximal and cannot increase without further cell division, would the cell cycle be the controlling factor for growth. The specific changes in the cell cycle that seem to be linked to flowering in *Silene* might suggest that changes in the expression of genes controlling development might depend on the cells being at some specific phase in the cell cycle but, so far, evidence on this is lacking.

5

The Subcellular and Biochemical Structure of the Meristem

5.1. Cytohistological zonation

The shoot apex can be differentiated into regions defined by morphology, by planes of division and by differences in growth rate and cell cycle. Can a corresponding biochemical differentiation of the cells also be detected and correlated with the cellular structure and the formation of initials and organs?

Sections of apices stained to show their general structure often show an intensity of staining which is least in the central zone, at the summit or centre of the apex, and greatest on the flanks of the apex. This zonation pattern is common to all types of vascular plants: ferns (Stevenson 1976), gymnosperms (Cecich, Lersten and Miksche 1972) and angiosperms (Nougarède 1967, Cutter 1971, Gifford and Corson 1971). The same pattern of staining is shown by general stains and by pyronin-Y, which is specific for RNA and therefore broadly indicates ribosome densities in the apex, since most RNA in the apex is in ribosomes (Lyndon 1977a). The zonation pattern is superimposed on the tunica–corpus structure and does not coincide with it (Lyndon 1968a, Molder and Owens 1973). Only rarely is there apparently no zonation, as in the *Lolium* apex, for example, where there is no indication of a central zone (Knox and Evans 1966).

The visibility of the central zone may also be enhanced by the nuclei being less heterochromatic and having more dispersed DNA (Rembur and Nougarède 1987). In the pea, the nuclei in the central zone are larger and more diffuse than nuclei with the same DNA content elsewhere in the apex. Nuclei with the 2C amount of DNA in the central zone are approximately twice the volume of 2C nuclei elsewhere in the pea apex (Lyndon 1973). This is probably because they are relatively more hydrated than nuclei elsewhere (Lyndon 1967).

5.1.1.   The molecular basis for zonation

The more general cytoplasmic stains that reveal zonation patterns tend to stain proteins and nucleic acids because these are the principal non-carbohydrate, insoluble cell components. Staining patterns for RNA are paralleled, but perhaps less strongly, by the patterns for protein. This may be because only a fraction of the cellular protein is in the form of ribosomes, and other proteins may not be so strongly zoned. In the pea apex there is about 9pg RNA per cell, corresponding to about 4.5pg of ribosomal protein, and since there is about 70pg of protein per cell, then about 94% is non-ribosomal (Lyndon 1970a).

Many authors have made the qualitative observation that the concentration of RNA in the central zone is less than in the peripheral zone on the flanks of the apex. This has been confirmed by quantitative and semi-quantitative measurements of RNA in shoot apices. In the moss *Polytrichum* (Hallet 1978) and the fern *Pteris* (Michaux-Ferrière 1981b) the apical cell and its immediate derivatives had a lower concentration of RNA than the cells on the flanks of the apex where leaves were initiated.

In higher plants, the peripheral zone stained more than the central zone in *Cosmos* (Molder and Owens 1972). The ratio of RNA concentration in the peripheral zone to that in the central zone was 1.5 in *Cannabis* (Heslop-Harrison and Heslop-Harrison 1970), 1.6 in *Chrysanthemum segetum* (Nougarède and Rembur 1976), and about 1.2 in 1-day-old *Chenopodium* plants, increasing to 1.5 in 6-day-old seedlings (Seidlová 1976). In *Pisum* the ratio (peripheral/central zone) was about 1.2, whereas for protein the ratio was less, *c*.1.05 (Lyndon 1970a). Similarly, in *Brachychiton* the ratio for RNA was 1.8, but for protein it was only

1.4 (West and Gunckel 1968b), indicating that the zonation pattern tends to depend more on distribution of RNA than of protein.

The concentrations of RNA are reflected in ribosome densities. A greater concentration of ribosomes in the peripheral zone than at the summit of the apical dome, as seen in *Pinus* (Cecich 1977), has been corroborated in angiosperms by direct counts of ribosome numbers from electron micrographs (Lin and Gifford 1976, Nougarède and Rembur 1976). In *Perilla* there were, per $5\mu m^2$, 3290 ribosomes in the central zone but 5115 in the peripheral zone of the apex, 1.55 times that in the central zone (Bronchart and Nougarède 1970).

These measurements give direct quantitative information about the zonation pattern. Although rates of incorporation of radioactively labelled RNA precursors have been used to get some measure of the relative rates of synthesis of RNA in the central and peripheral zones, these must be treated with great caution. The rate of incorporation of a precursor will depend not only on the rate of synthesis of the end-product but also on precursor pool sizes in the cells and compartmentation of exogenous and endogenous precursors, and possible metabolism of applied precursors into compounds not eventually measured. In wheat, the ratio (peripheral zone/central zone) of $[^3H]$uridine incorporation was about 1.3 (Evans and Berg 1972b). In *Brachychiton* the rate of incorporation of $[^3H]$orotic acid into RNA was in the ratio (peripheral zone/central zone) of 1.5, but this difference was less than expected, about 1.8, on the basis of the relative RNA concentrations (West and Gunckel 1968b). However, in the *Euphorbia* apex, despite the mitotic index and therefore probably the division and growth rates in the peripheral zone being twice that in the central zone (Raju 1968), the incorporation of label into RNA was uniform throughout the apex (Raju and Ho 1973). Also, in *Chenopodium* (Seidlová 1976) and *Pisum* (Lyndon 1972a) the incorporation into the peripheral and central zones was equal, despite a difference in RNA concentrations having been demonstrated in both cases quantitatively or semi-quantitatively. This was interpreted as showing in the pea that probably the pool size of precursors was smaller in the central zone than in the peripheral zone so that the incorporation rates were not a true measure of the rates of RNA synthesis (Lyndon 1972a).

The zonation pattern is therefore a reflection of the generally higher concentrations of RNA (and protein) on the flanks of the apex, in the peripheral zone, than in the central zone (Fig. 5.1). But this does not necessarily reflect the amounts of RNA per cell. In *Pisum* and *Silene*, mean cell number per unit volume was lower, and so cell volume was

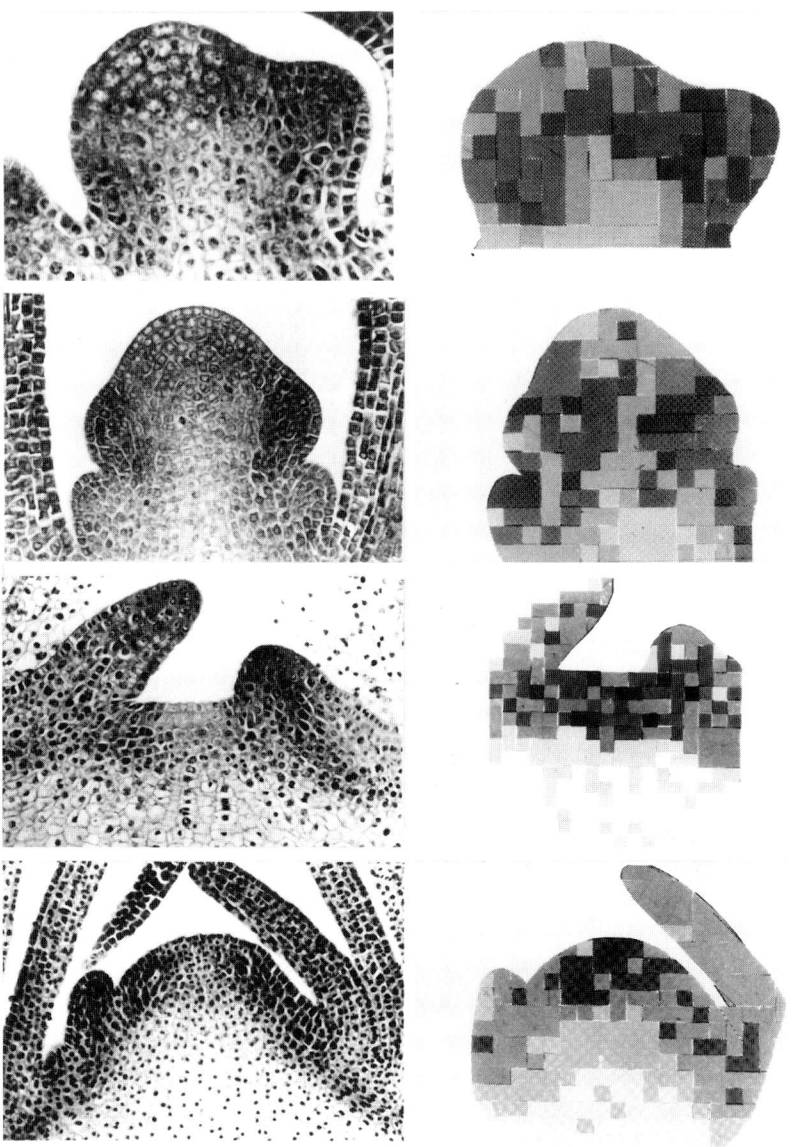

Figure 5.1 Shoot apices (L.S.) of (top to bottom) *Pisum sativum*, *Silene coeli-rosa*, *Helianthus tuberosus* and *Chrysanthemum uliginosum* stained with pyronin-Y to show RNA distribution (left) and shaded according to cell densities (right). The two patterns are similar in *Pisum* and *Silene* where cytohistological zonation corresponds to cell density, and RNA per cell is therefore similar throughout the apical dome and in the primordia. However, the patterns are dissimilar in *Helianthus* and *Chrysanthemum*, showing that the smaller cells of the central zone (higher cell density) have less RNA per cell than the cells in the primordia. From Lyndon (1972b).

larger, in the central zone than in the peripheral zone. The zonation pattern and distribution of RNA were therefore directly related to cell volume and inversely related to cell number per unit volume (Lyndon 1972b). In *Pisum*, the ratio of RNA concentration in the peripheral zone to that in the central zone was 1.2:1 (Lyndon 1970a), the inverse of the ratio of cell volumes, which was 1:1.2 (Lyndon 1968a). This was consistent with RNA per cell being constant throughout the apex and RNA concentration therefore varying inversely with cell size (Lyndon 1970a). In *Pisum* the measured amounts of protein and RNA per cell were the same in the central zone and the peripheral zone but, in the central zone, the cells were larger (Lyndon 1968a), therefore their contents were less dense (Lyndon 1970a). *Chrysanthemum segetum* is similar (Nougarède and Rembur 1976). The converse picture was found in *Helianthus tuberosus* and *Chrysanthemum uliginosum*, in which the cells tended to be smaller (and so more numerous per unit volume) in the central zone, implying that RNA per cell was least in this region, and greater in the peripheral zone (Lyndon 1972b). This sort of information, of the relation of cell number to zonation pattern, does not require the use of sophisticated apparatus but can be deduced from the zonation pattern and a knowledge of the number of cells per unit volume in the various regions of the apex (Lyndon 1972b).

These observations mean that in some plants, such as *Chrysanthemum segetum*, *Pisum* and *Silene*, the zonation pattern in the apical dome is mirrored by cell volume, whereas in other plants (even, in *Chrysanthemum*, in the same genus) it is not. However, in the pith–rib meristem, where there is considerable cell enlargement, the vacuolation there results in lower concentrations per unit tissue volume of all cytoplasmic constituents and so contributes to the zonation pattern.

The zonation pattern that seems common to many vegetative apices therefore depends on the concentration of RNA (and to a lesser extent, protein) per unit volume of tissue, irrespective of the number of cells into which the tissues are divided. There is no general congruence of zonation with cellular structure: the concentrations of protein and RNA seen as zonation seem to be independent of the underlying cell sizes and cell densities (Lyndon 1972b). The compartmentalization into cells, therefore, seems not to be relevant for whatever aspect of apical functioning the zonation pattern may be essential. The virtual ubiquity of the zonation, but not the cellular structure, suggests that the zonation pattern is in some way linked to apical functioning, especially the formation of leaves and the absence of primordia from the surface of the apical dome, and that the cellular structure is more the consequence of apical growth than its cause.

5.1.2.   Nucleolar sizes

Differences in nucleolar size associated with zonation suggest that there might also be differences in nucleolar functioning. Actively growing cells often tend to have larger nucleoli than slower growing cells, since the size of the nucleolus reflects its activity (Alberts et al. 1994). Slowly growing cells may also have smaller nucleoli because G_1 would then tend to be the predominant phase of the cell cycle and so the nucleoli will be just at the beginning of their growth during the cell cycle. Nucleoli are smaller in the central zone in *Aster*, *Chrysanthemum* and *Papaver*, but not in *Chenopodium*, in which the nucleoli in the central zone were often larger than in the peripheral zone (although this may be because these were intermediate meristems, i.e. progressing towards flowering, rather than being strictly vegetative) (Nougarède 1967). In *Amaranthus* there was no difference in nucleolar volumes between central and peripheral zones in vegetative apices, but in apices in transition to flowering nucleoli in the peripheral zone tended to be larger than in the central zone at maximal area phase (Nougarède 1967). But in *Amaranthus*, even where no obvious differences in nucleolar volumes could be found, there was still a zonation with lower RNA concentration in the central zone (Nougarède, Gifford and Rondet 1965). In *Cosmos* nucleolar volumes in the shoot apex changed during development but not in parallel with RNA content (Molder and Owens 1973).

The conclusion must be that although much of the zonation is because of lesser RNA and ribosome concentrations at the summit of the apex, this is not always paralleled by differences in nucleolar size. The dynamics of RNA and ribosome synthesis and turnover in relation to changes in nucleolar size are not yet sufficiently understood to allow us to relate them to apical zonation and functioning in any helpful fashion.

5.1.3.   Changes in zonation pattern

The zonation pattern is not constant; it may vary according to the stage of the plastochron in vegetative apices and may become modified or disappear on the transition to flowering. In the vegetative apex, the central zone tends to diminish in size, or even temporarily disappear, at the minimal area stage of the plastochron at about the time of primordium initiation, as shown in *Pisum* (Lyndon 1968a) and *Cosmos* (Molder and Owens 1972). In both *Pisum* and *Cosmos* there was no change in RNA concentration in the central zone relative to the peripheral zone during the plastochron. However, when the central zone became less evident, in *Cosmos* its cells, and in *Pisum* its nuclei, tended to be smaller. In *Pisum* the cell cycle in the central zone is about 1.4 times as long as the plasto-chron (Lyndon 1970b) and so the changes in the size of the central zone must reflect reversible changes in cells while they are progressing through a single cell cycle.

On the transition to flowering, the distribution of RNA tends to become more uniform throughout the apex (Nougarède 1967) and the zonation pattern tends to disappear, as in *Nicotiana* (Thomas and Kanchanapoom 1990) and *Cosmos* (Molder and Owens 1973). In *Brassica campestris* the zonation also tends to disappear in the meristem in transition to flowering, but is re-established in the inflorescence apex but not in the meristems of the individual flower meristems (Orr 1978). Zonation is also re-established in *Sinapis* (Bernier 1969) but is never lost in *Impatiens* (Simon 1973), both plants in which the flowering apex can revert back to vegetative growth (Battey and Lyndon 1990). Whether the disappearance of zonation implies a more complete commitment to flow-ering of the apex remains to be resolved.

But how is the zonation pattern determined and maintained, and what is the physiological basis underlying it? Is the zonation pattern a reflec-tion of cell cycle lengths, i.e. is there a higher concentration of RNA in faster growing cells? For the pea there are detailed measurements for both RNA concentrations and for cell cycle lengths and for distribution of mitoses. The region of the apical dome in which cell division is slower is much larger than the central zone in which RNA concentrations are lower. There is no exact parallel between RNA distribution and cellular growth and division rates (compare Fig. 5.1a with Fig. 4.1) (Lyndon 1970a, b, 1972b). The zonation pattern in the pea apex is therefore not simply a visible manifestation of the pattern of growth rates and cell cycle lengths.

Is there perhaps a gradient of growth substances or metabolites that determine the zonation pattern in the shoot apex or something present in the central zone which inhibits growth and RNA, ribosome and protein synthesis in its cells? Why does the zonation pattern vary during the plastochron; is it influenced by signals from the young primordia? So far, lack of data means that these questions remain unanswered and the significance of the zonation pattern eludes us.

5.2.   Biochemical activity in the shoot apex

Because it is so tiny, and the consequent difficulty of obtaining tissue and of making measurements, most of the information about metabolic activity in the shoot apex has been provided by histochemical investigations on sections. However, direct assays of biochemical activity have also been made by micromethods which allow measurements to be made on pieces of tissue only a fraction of a mm^3 in size. Most of the measurements within the shoot apex have been of the activities of enzymes concerned with respiratory metabolism.

5.2.1.   Respiratory enzymes

Succinic dehydrogenase and cytochrome oxidase are key enzymes, respectively, of glycolysis and electron transfer, and have been demonstrated histochemically in a number of shoot apices. Succinic dehydrogenase and cytochrome oxidase activity are usually found throughout the apical meristem (*Rauwolfia*, Mia and Pathak 1968) but often tend to be greatest at the summit of the apex and in the young leaf primordia (*Pinus*, Fosket and Miksche 1966; *Triticum*, Evans and Berg 1972a; *Brassica campestris*, Petersen and Orr 1983, Orr 1984; *Saccharum*, Thielke 1965; *Pharbitis*, von Klopfer 1973). In contrast, in *Sinapis* (Jacqmard 1978) and in very young *Pinus* seedlings (Fosket and Miksche 1966) activity of succinic dehydrogenase and cytochrome oxidase respectively were greatest on the flanks of the apex. The tunica may sometimes differ from the corpus cells: cytochrome oxidase activity was least in the tunica in *Saccharum* (Thielke 1965) and *Xanthium* and greatest in *Pharbitis* (von Klopfer 1973).

Other enzymes associated with respiratory pathways have been demonstrated histochemically in shoot apices. In wheat apices, malate dehydrogenase occurred throughout the apex (Opatrná 1975), alcohol dehydrogenase and isocitric dehydrogenase were concentrated at the tip of the shoot apex (Opatrná 1974, 1975) and lactic dehydrogenase became detectable only as the apices aged (Opatrná 1975). All these enzyme activities increased on the transition to reproductive growth, and their distribution became more homogeneous (Opatrná 1970, 1975). The presence of these enzymes is consistent with the operation of the tricarboxylic acid (TCA) cycle in respiration in shoot apices.

The activities of glucose 6-phosphate dehydrogenase and 6-phospho-gluconate dehydrogenase, enzymes of the pentose phosphate pathway (PPP), were also examined in *Brassica*. Like succinic dehydrogenase and cytochrome oxidase, activities were highest in the vegetative meristem, in the central zone. At floral transition they increased in the peripheral zone to match the central zone and at late transition the peripheral zone especially showed high activity, as did the floral buds (Orr 1985).

Phosphofructokinase is a key enzyme of the glycolytic pathway of respiration and glucose 6-phosphate dehydrogenase is a key enzyme of the PPP, and their activities have been taken to represent the relative activities of the glycolysis pathway and PPP respectively. Measurements by the sensitive micromethod used by Croxdale and Outlaw (1983) showed that, in *Dianthus*, phosphofructokinase activity was the same in the apical meristem and in the youngest leaf primordium, but declined in the primordia as they developed (Croxdale 1983). Glucose 6-phosphate dehydrogenase activity in the youngest leaf primordium was about double that in the apical meristem but, as the primordia developed, this activity, too, declined until after nine plastochrons it was again the same level as in the apical meristem (Fig. 5.2). In the apex the activity of phosphofructokinase (1056 mmol kg^{-1} dry wt h^{-1}) was four times that of glucose 6-phosphate dehydrogenase (250 mmol kg^{-1} dry wt h^{-1}). The capacity for glycolysis therefore apparently exceeded that for the PPP in the apex and primordia but became relatively less important, and the PPP relatively more important, in the maturing regions (Croxdale 1983). Activity of hexokinase (another glycolytic enzyme) was also highest in the apex, slightly less in the youngest leaf primordium and declined with primordium development (Croxdale and Vanderveer 1986). In leaves that were older but still growing, hexokinase activity was least at the tip of the leaf, where the cells are first to mature, and greatest at the base, where growth continues longest. This decreasing hexokinase activity with increasing maturation of the leaf tissues is consistent with a decline in glycolytic activity as cells matured.

It can be summarized what seems to be the picture in the shoot apex. Respiration seems to be greatest at the apical summit in young apices but in older apices it may be greater in the deeper, central cells. There may be little difference in respiratory activity between the central zone and the flanks of the apex and the young leaf primordia. The glycolysis pathway

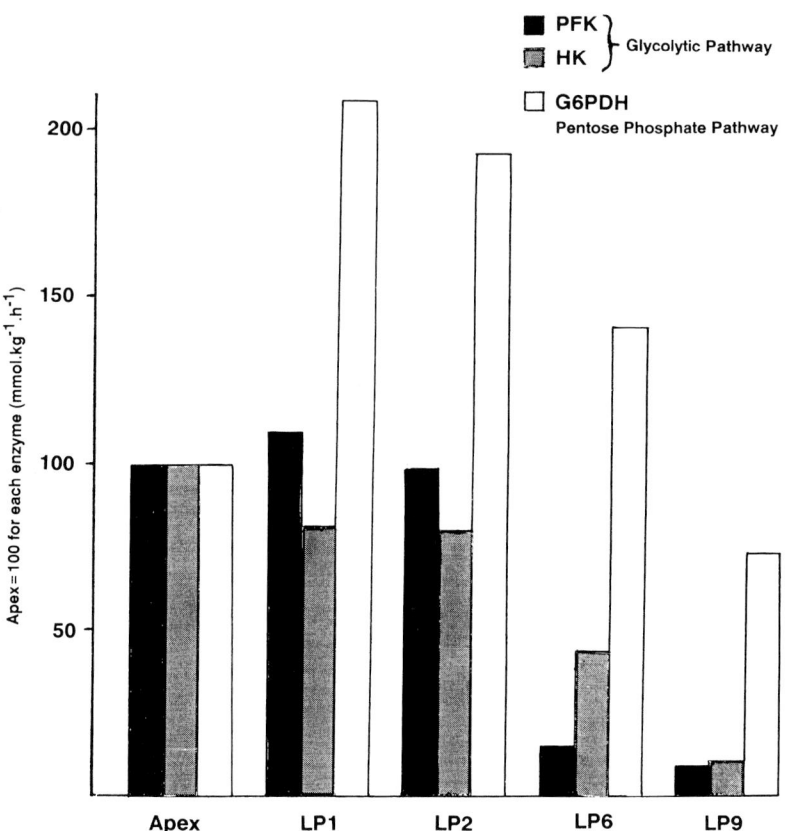

Figure 5.2 Respiratory patterns in *Dianthus* leaf primordia as shown by the relative activities of the enzymes of the glycolytic pathway (phosphofructokinase, PFK; and hexokinase, HK) and the pentose phosphate pathway (PPP) (glucose 6-phosphate dehydrogenase, G6PDH). Relative to the shoot apex ($=100$) the PPP becomes more important and glycolysis less important as the primordia develop from leaf primordium 1 (LP1) to leaf primordium 9 (LP9). Data from Croxdale (1983) and Croxdale and Vanderveer (1986).

seems to predominate at the apex and in the youngest leaf primordia but the PPP may become more predominant as the organs, and tissues produced by the meristem, mature.

However, in comparing the relative enzyme activities in different parts of the same apex, caution is needed. Even if differences in enzyme activity can be shown, and are valid, these are not necessarily indications of differences in respiration rates of the cells, nor does the occurrence of a particular set of enzymes preclude respiration proceeding by another pathway. Although the relative activities of phosphofructokinase and glucose 6-phosphate dehydrogenase are thought to be indicative of the the relative importance of the glycolysis and PPP pathways of respiration, respectively, the activities of single enzymes are unlikely to be simply correlated with the rate of respiration of the whole cell or the pathway by which it takes place. All that the differential enzyme activities can tell us is the potential for a particular metabolic pathway involving that particular enzyme.

5.2.2. *Rates of respiration*

Rates of respiration of isolated shoot apices have been measured directly by the use of a Cartesian diver microrespirometer technique (Ball and Boell 1944). Detailed measurements were made on apical fragments excised from shoot tips of *Lupinus* seedlings (Sunderland, Heyes and Brown 1957). Each consisted only of the apical dome, or the apical dome plus one or more leaf primordia and their associated stem tissue. From comparable fragments the primordia were each carefully excised for measurement. The values for a given apical unit (primordium plus stem tissue) were obtained as the differences between one apical fragment and the next and the values for internode (plus node) were obtained by difference between the value for the appropriate apical unit and the value for the associated primordium. Thus, all measurements were made on a piece of tissue with only a single cut surface. The protein content and cell number of each of the fragments was also measured.

In the primordia, the rate of respiration (O_2 uptake h^{-1}) increased exponentially with primordium serial number, and to a greater extent than cell number, which also increased exponentially (Sunderland and Brown 1956), so that respiration per cell increased linearly with increasing primordium plastochron age (Fig. 5.3). In the internodes, respiration per cell did not increase as it did in the primordia but respiration per cell in the first (youngest) internode was already as high as, or higher than, in primordium 7. Since the primordia were assumed to be primarily derived from tunica cells, and the internode from corpus, it was suggested that the tunica showed a low respiration rate and the corpus a higher rate, perhaps as much as five times greater on a per cell or a per unit protein basis. The primordia at initiation, far from being active growth centres, indeed seemed to be the less active parts of the meristem as judged by respiratory rate.

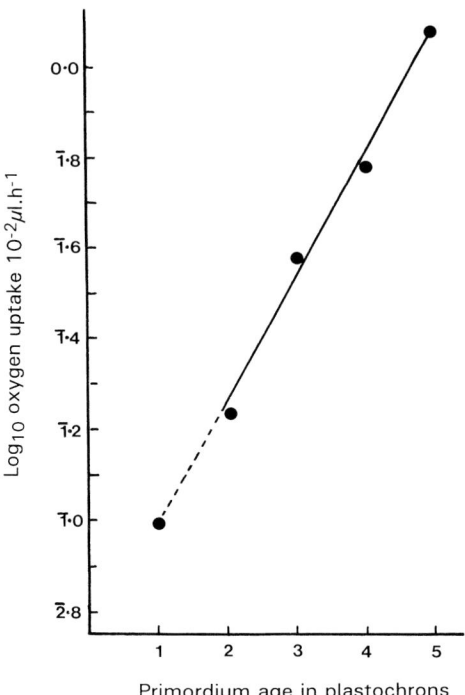

Figure 5.3 The rate of oxygen uptake of *Lupinus* leaf primordia increased exponentially with plastochron age. Data from Sunderland, Heyes and Brown (1957).

These results for *Lupinus*, showing respiration is probably less active in the tunica of the apical dome than in young developing primordia, are consistent with these apices being from plants in which the tunica–corpus structure is well developed and where respiratory activity is greatest in the subterminal cells of the meristem (as was the case for succinic dehydrogenase and cytochrome oxidase activities). The overall picture then is that, in general, respiratory activity is probably fairly evenly distributed throughout the apical meristem when the seedling is young. However, when a distinct tunica becomes established this becomes less active than the cells below it in the corpus, and the cells on the flanks of the apex. Also, the apical meristem is more active than the young primordia it produces. Only after they have grown for seven or so plastochrons do the primordia 'catch up' with the apex.

This is in contradiction to the interpretation of Croxdale (1983) from enzyme activities in *Dianthus*, that the respiratory activity of the primordia declines as they age. Either *Lupinus* and *Dianthus* show opposite trends and so there is no general trend, or else different methods give different results. Where the actual rate of respiration is concerned, it would seem preferable to believe the direct results of Sunderland, Heyes and Brown (1957). Histochemical methods and the methods of Croxdale (1983) may be more appropriate for indicating possible relative contributions of different pathways rather than absolute rates of respiration. If this is so, then the apical dome may be more active than the young primordia in respiration, which in the apical dome may be primarily by glycolysis but as the cells produced by the meristem develop, the PPP becomes more important. This may be related to the formation of vascular tissues which have been shown, in the root, to respire predominantly by the PPP whereas in the cortical cells the glycolysis pathway is more important (Fowler and ap Rees 1970).

On flowering, in the wheat apex the respiration rate increased from 5.6 to 8.1 ($10^{-3}\mu$l h^{-1} μg^{-1}DW), as shown by direct microrespirometric measurements (Krekule and Teltscherova 1966) and this may be linked to the increase in mitochondria which occurs in the apex on flowering (see Chapter 9).

The pattern of respiratory activity, as measured directly and by histochemistry, does not seem to be the same as the cytohistological zonation and sometimes seems to be more related to the tunica–corpus cellular structure. Moreover, it does not seem to be directly related to the growth rate or the cell cycle of the cells, except when the apex becomes floral. We seem to have at least three different patterns superimposed on each other in the vegetative shoot apex: cellular structure, cytohistological zonation and respiratory activity.

5.2.3. *Other enzymes and proteins*

In very young *Pinus* apices, acid phosphatase activity was high at the extreme tip of the shoot meristem, but in older apices it was essentially confined to the peripheral zone where leaves were initiated (Fosket and Miksche 1966), as in wheat (Evans and Berg 1972a) and *Sinapis* (Jacqmard 1978). In *Rauwolfia* acid phosphatase was not detected in the apical meristem although it was present in more basal differentiating cells (Mia and Pathak 1968). In the *Sinapis* apex on flowering, acid phosphatase, ribonuclease and succinic dehydrogenase tended to become more intensely and evenly distributed throughout the apical dome. The increases in these enzyme activities corresponded, respectively, with the dispersal of vacuolar material (14h after the beginning of floral induction), just after a rise in RNA synthesis (18h), and simultaneously with an increase in mitochondria (22h) (Jacqmard 1978). The changes in numbers of organelles, with which the enzyme changes were coincident, may be more easily quantified and could suggest that the measured enzyme activities are simply a function of the amounts of enzymes present (as would perhaps be expected) and do not necessarily indicate actual *in vivo* activities of the pathways to which they contribute.

Adenylate cyclase (involved in cAMP synthesis) has been demonstrated in shoot apical cells of the moss *Bryum* (Bhatla and Chopra 1984).

5.2.4. Metabolites

Sucrose has been measured throughout the apical meristem of the *Sinapis* apex by the sensitive micromethods of Bodson and Outlaw (1985). The concentrations of sucrose and soluble sugars in the central zone and the peripheral zone were the same, although they increased by $> 50\%$ when the meristems were evoked to flower. These authors suggest that 'factors other than those related to the soluble carbohydrate metabolism are responsible for meristem zonation' (Bodson and Outlaw 1985, p.423).

In the vegetative apex, starch grains are found principally in the cells, at the base of the apex, which form the incipient pith, in wheat (Evans and Berg 1972a), *Brachychiton* (West and Gunckel 1968a), *Chenopodium* (Seidlová 1977) and *Pisum* (Lyndon and Robertson 1976). It was also the tissues basal to the meristem in which phosphorylase (concerned with starch synthesis) was found in *Rauwolfia* (Mia and Pathak 1968). Although not always evident by histochemical staining in the light microscope, electron microscopy reveals small starch grains throughout the pea apical meristem (Lyndon and Robertson 1976). In *Chenopodium* there was essentially no difference in starch distribution in vegetative, induced and prefloral apices, suggesting that the presence of starch was linked with organogenesis rather than 'with the quantitative aspects of growth' (Seidlová 1977, p.389). Changes in starch distribution were, in fact, closely linked to leaf initiation in the pea (Lyndon and Robertson 1976) (see Chapter 6). Although the presence of starch sometimes seems to be linked in some way to organogenesis, particularly shoot formation (Thorpe and Murashige 1970), the significance of starch accumulation in the shoot apex and its implications for changes in sucrose concentrations and carbohydrate metabolism are not yet properly understood.

Cell wall material was measured by differential extraction procedures combined with periodic acid-Schiff staining in shoot apices of *Brachychiton* (West and Gunckel 1968a). In the central zone of the apex, pectic substances were the main wall constituent. As the cells developed into the pith, non-cellulosic polysaccharides and to a smaller extent hemicellulose increased. Cellulose was only a minor constituent of the walls (< 3%) at all stages of development measured. It may be a more prominent constituent of the outer epidermal walls where it can be detected by polarized light microscopy (see Chapter 7).

When the apical cell and adjacent cells, mostly the relatively large cells of the outer vacuolated layer of the shoot apex, of the fern *Adiantum* were dissected out and analysed, 72% of the free nitrogenous compounds was found to be in the form of a single compound – γ-hydroxy-γ-methylglutamic acid, compared with 57% in the older leaves. The proportion was even higher, 91%, in the young leaf primordia. Similarly, in amino acid composition of their protein, the shoot apex cells differed more from the young leaf primordia than from the older leaves. It was concluded that 'the large, vacuolated outermost cells of the apex, including the apical cell itself, are different from the smaller, denser, dividing cells below, and that they resemble more closely, in their composition, older tissue in which divisions have virtually ceased' (Steward, Wetmore and Pollard 1955, p.948). This is similar to the conclusions drawn about the distribution of respiratory activity (see Sections 5.2.1, 5.2.2).

This fragmentary information about the occurrence and distribution of metabolites in the shoot apex so far shows only that the biochemical activities within the apex may be largely similar to those of adjacent tissues, and other organogenetic regions in the plant.

5.3. Ultrastructure of meristems

The ultrastructure of vegetative shoot meristem cells has likewise shown nothing especially distinctive about them. Vacuoles are often small except in the developing cells of the pith-rib meristem (Gifford and Stewart 1967). Even in *Helianthus*, which has a well-marked central zone, no differences could be found between cells of the central zone and the rest of the meristem, except that there were starch-containing plastids in the central zone but not generally in the peripheral zone (Sawhney, Rennie and Steeves 1981). As in the pea meristem, this indicates a metabolic rather than an ultrastructural difference. The only consistent difference between the cells of the meristem is a greater concentration of ribosomes in the peripheral zone than in the central zone (Nougarède 1967, Lin and Gifford 1976), consistent with the zonation as seen by staining (see Section 5.1). In *Glechoma* the epidermal cells stained more densely than the rest of the apical cells (Bowes 1965), and occasional reports suggest a difference between the tunica and the corpus, although all cells seem to be interconnected by plasmodesmata (Lynch and Rivera 1981). In the central zone of the *Helianthus* meristem there was starch in the plastids only of the two tunica layers (Sawhney, Rennie and Steeves 1981).

A detailed analysis of the relative volumes occupied by various organelles was made for some cactus shoot apical meristems. In the young and in the mature *Echinocereus* meristem there were no apparent differences between different regions of the apex, nor between tunica and corpus, nor between cells as they were displaced down the apex within a region, nor between different plastochron stages, nor between the apex at germination and the apex of the mature shoot, except for a little more vacuolation in the mature apex (Mauseth 1980, 1981a, b, 1982a). The shoot apical meristem of another cactus, *Trichocereus*, had a similar ultrastructure to that of *Echinocereus*, except for smaller plastid and dictyosome volume (Mauseth 1982b).

Although the quantitative structural differences between different parts of the same meristem and meristems of different species and the same species under different growth conditions were statistically significantly different, there were no clear correlations of structure with any other parameter or growth rate. Even when meristem size was doubled by treatment with benzylaminopurine, the structure of the *Opuntia* meristem was unchanged (Mauseth 1976). The overriding impression is that the quantitative ultrastructure, on a volume basis, of the different regions of the cactus meristems is fairly similar.

A comparison of the quantitative ultrastructure of the meristems of two cacti and of *Pisum* and *Sinapis* showed that although there were differences in the general organelle composition, there was as much difference between the two cacti as between the cacti and *Pisum* and *Sinapis* (Table 5.1) (Mauseth 1982b). Pea apex cells were essentially non-vacuolate. The cells were much more vacuolate in the cacti, and there were lesser amounts of dictyosomes than in *Pisum*. All these points would be consistent with the cacti having less intense meristematic activity, and the cacti and *Sinapis* meristems having a slower growth rate than the pea (see Table 3.4, Chapter 3). When the vacuole is discounted, the

Table 5.1. *Comparison of quantitative ultrastructure of the shoot apices of two cacti,* Echinocereus *and* Trichocereus, *with* Pisum *and* Sinapis.

Percentage of protoplast volume **or protoplast volume minus vacuole**							
	Echinocereus		*Trichocereus*		*Pisum*		*Sinapis*
Vacuole	21.4		24.2		0.39		11.1
Nucleus	29.7	**37.6**	22.2	**29.2**	40.5	**40.5**	33.7	**37.8**
Mitochondria	6.4	**8.1**	6.9	**9.0**	5.3	**5.3**	3.2	**3.6**
Plastid	5.8	**7.3**	2.7	**3.6**	5.6	**5.6**	4.5	**5.1**
Dictyosomes	0.4	**0.5**	0.2	**0.2**	1.3	**1.3**	<0.1	**<0.1**
Hyaloplasm	36.4	**46.1**	43.8	**57.6**	45.8	**45.8**	48.1	**54.0**

Data from Mauseth (1982c).

percentage of protoplast volume occupied by the nucleus and the cytoplasmic organelles is quite similar in the cactus and the pea but there was less mitochondrial matter in pea and mustard.

In the cacti, the main changes in ultrastructure were seen as the cells, and organs, leaves and spines, began to differentiate. Leaf and spine primordia were less vacuolate than the apical cells but showed essentially no increase in plastid content (Mauseth 1982c), presumably because the leaves especially are very reduced in the cacti. However, in the developing cortical chlorenchyma, the main photosynthetic tissue in the cactus stem, there was a major development of chloroplasts (Mauseth 1981b). Similarly, in *Glechoma*, the proplastids were least developed in the cells of the initiating zone, just under the epidermis, and the chloroplast lamellae developed as the cells became more basal. In the outer flank, many young chloroplasts could be seen and the cells became much more vacuolated (Bowes 1965).

Correlations of ultrastructure relative volumes with metabolic rates, etc., may not be very helpful because for reactions such as respiration involving surfaces the number of organelles (or profiles as seen in sections) may be a much more useful indicator of cellular activity and differentiation. Although such values are not available for all the detailed work on cacti, they are available for *Pisum* and *Sinapis* meristems. In the *Pisum* apex, the relative volume of the cellular components did not appreciably alter throughout the meristem except for an increase in plastid and vacuolar volume and an increase of endoplasmic reticulum in the rapidly enlarging cells of the incipient pith (Lyndon and Robertson 1976). However, the numbers of organelles did change (Fig. 5.4). The number of plastids increased in the young leaf primordium, as did the

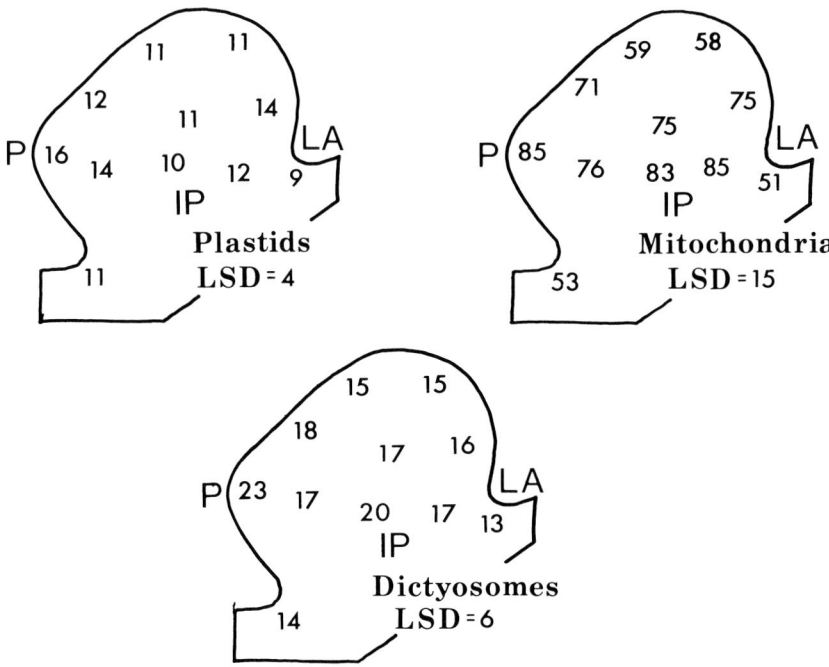

Figure 5.4 Numbers of plastids, mitochondria and dictyosomes per cell in the indicated regions of the *Pisum* shoot apex (L.S.). P, youngest primordium; LA, leaf axil; IP, incipient pith; and LSD, least significant difference (*p* = 0.05). From Lyndon and Robertson (1976).

number of mitochondria and dictyosomes. In the incipient pith, despite their increase in size, the number of plastids per cell remained constant, but numbers of mitochondria and microbodies (e.g. glyoxysomes, etc.) increased. There were fewest plastids, mitochondria and dictyosomes per cell in the developing leaf axils, where the axillary buds would eventually form. The conclusion was that ultrastructural changes are linked to cell and organ differentiation and not to plastochronic functioning or division rate. In *Sinapis* the main ultrastructural changes occur at the developmental switch to flowering, when the number of mitochondria increases and is probably linked to an increase in respiration and metabolic rate at this time (Havelange, Bernier and Jacqmard 1974). A similar increase in mitochondria also occurred when *Sinapis* plants were given high irradiance without inducing flowering (Havelange and Bernier 1983).

In *Arabidopsis*, a decrease in the amount of DNA per mitochondrion from > 1Mbp in the shoot apical meristem to 170kbp in the mature foliage leaf was apparently because DNA synthesis did not keep up with mitochondrial division (Fujie et al. 1994) and the mitochondria became smaller, as they did in the young leaf primordium in the pea shoot apex (Lyndon and Robertson 1976). On the other hand, DNA per plastid in *Arabidopsis* increased from about 1 to 4Mbp in the shoot apex to >90Mbp in the mature leaf as the plastids multiplied and enlarged (Fujie et al. 1994).

In pea, the ultrastructural changes were correlated only with cell differentiation and not with plastochronic functioning or growth and division rates throughout the apex rate. It was salutary to find that apparently obvious differences in ultrastructure between different regions of the apex which caught the eye were not borne out by objective quantitative analysis, and yet the most striking difference – the increase in the number of plastids containing starch as the plastochron progressed (see Chapter 6) – was not noticed until the quantitative analysis had been done. So much for visual analysis of electron micrographs!

5.4.   Gene expression

The minute size of the shoot meristem means that only vanishingly small amounts of tissue are available for molecular and biochemical analysis. This has been overcome by using cauliflower heads, very highly branched shoots which are a collection of thousands of shoot meristems. By means of the powerful polymerase chain reaction (PCR) technique, some cDNAs were isolated which were expressed in the shoot meristem, but not the leaves, of transgenic *Arabidopsis* (Medford, Elmer and Klee 1991). By fusing the promoter region of one of these cDNAs, *meri-5*, to the GUS reporter gene, it was shown that *meri-5* was not only most strongly expressed at the branching points in the stem but also in the shoot meristem and young primordia in both vegetative and floral apices. It was also expressed, but only at lower levels, in the root meristem, the vascular tissues (especially the internal phloem), and the receptacle and pedicel of the flower. Another gene, *meri-1*, codes for a histone, H3, and its promoter region directed GUS expression in transgenic tobacco particularly on the flanks of the shoot apex, where primordia are formed, but not in the central zone at the summit of the apex where the initial cells are found. It also was expressed strongly in the procambium (Medford, Elmer and Klee 1991). In transgenic tobacco plants, the *PCNA-GUS* chimeric gene was shown to be similarly expressed, in the shoot apical meristem and young primordia and in the meristem of the root (*PCNA* is proliferating cell nuclear antigen, a gene essential for DNA replication) (Kosugi et al. 1991). The *THOM1* homeobox gene in tomato is also strongly expressed in the shoot apical meristem and the procambium (Meissner and Theres 1995). These same sorts of distribution are like those often found for general nucleic acid and protein stains and probably serve to indicate the regions of small, dense and dividing cells, rather than anything more specific. Another gene expressed in the peripheral zone, but not the central zone, of the apex was that for the storage protein cruciferin in the *Brassica napus* embryo apex. The gene for napin, the other main storage protein, was expressed throughout the apex (Fernandez, Turner and Crouch 1991).

Other genes that are expressed specifically in meristematic cells are five cDNA clones showing homology to histone genes, and are presumably linked to cell division activity (Köhler et al. 1992). The gene for plant histone H2A is expressed throughout tomato shoot apices, but shows a spotty distribution, being very reminiscent of an apex labelled to highlight DNA synthesis (Konig et al. 1991). This strongly suggests that this gene is cell cycle regulated and may be expressed during late G_1 and S-phase. Expression patterns of five cell cycle-related genes in *Antirrhinum* also showed this type of dispersed distribution, here clearly shown to be limited to individual cells, and consistent with the cell cycle being asynchronous in this meristem (Fobert et al. 1994). Another two cDNA clones, for ribosomal genes, also showed a spotty distribution (Köhler et al. 1992), again suggesting that these genes are also being expressed only at some specific stage of the cell cycle. This reminds us how little is known about the kinetics of synthesis and release of ribosomes in relation to the cell cycle.

Two genes with complementary expression patterns have been found in tomato. A polyphenoloxidase gene is expressed in the vegetative shoot meristem, but strongly only in the epidermis, and in floral primordia but not in the apical meristem of the flower (Shahar et al. 1992). A complementary distribution is shown by a gene coding for a dUTPase. This gene is expressed in the vegetative shoot meristem, although not in the epidermis, but not in the flower meristem except for the flower primordia themselves where it is expressed in the epidermis and L3 but not in L2 (Pri-Hadash, Hareven and Lifschitz 1992). Other genes with expression in the vegetative shoot meristem include two cDNA transcripts confined to the epidermis in the succulent *Pachyphytum* (Clark, Verbeke and Bohnert 1992), and a lipid transfer protein gene in *Nicotiana* which shows high expression in the epidermis, lower in the underlying cell layers, and again higher in the submeristem region (Fleming et al. 1992). Genes expressed in some cell layers and not others confirm the layered nature of floral as well as vegetative meristems but have yet to throw light on the functional and morphogenetic differences between layers.

The *knotted1* and related genes are particularly interesting. *KN1* is expressed in the corpus of the shoot apex of maize, but not at the P_0 leaf initiation site, where the primordium is just a slight bulge and periclinal divisions are visible in the tunica (i.e. the epidermis in maize) (Jackson, Roberts and Martin 1992). Interestingly, the gene product, the KN1 protein, is expressed in the epidermis as well, suggesting that gene expression is mediated by a diffusible gene-produced factor. The related genes *RS1* and *KNOX3* are expressed in a ring of internodal initial cells and in the axillary bud. All four homeobox genes (*KN1*, *RS1*, *KNOX3* and *KNOX8*) are expressed evenly throughout the corpus of the inflorescence meristem and in the spikelet and flower meristems but not in lateral organs such as glume primordia. These genes seem to be concerned with preserving meristematic capability since they are expressed only in indeterminate meristems. As soon as the meristem cells begin to contribute to a determinate organ, expression of these genes is suppressed. A *knotted1*-like homeobox gene in *Arabidopsis* is expressed in the vegetative shoot apical meristem primarily in the peripheral zone and not in the central zone, again showing a correlation with more intense meristematic activity, but when overexpressed in transgenic plants it causes severe lobing of the leaves, presumably by prolonging indeterminate meristematic activity in the leaf (Lincoln et al. 1994).

A cDNA library was constructed from RNA extracted from shoot apical meristems (the apical dome and the youngest primordium) dissected out from tomato shoots. Some of these cDNAs, randomly selected, were then used for *in situ* hybridizations in longitudinal sections of apices (Fleming et al. 1993). None of the genes was restricted in expression to the shoot apical meristem – they were also expressed elsewhere in the plant. However, six different patterns of expression in the shoot apical meristem were found (Fig. 5.5; Table 5.2). The gene for lipid transfer protein was expressed only in the epidermis, and histone genes were expressed with the spotty distribution found by other workers (see above). The dUTPase gene was expressed particularly in the corpus. The small subunit of Rubisco was expressed only in the young primordia and not in the meristem, being consistent with the development and differentiation of the plastids in the incipient leaves (compare Fig. 5.4). The

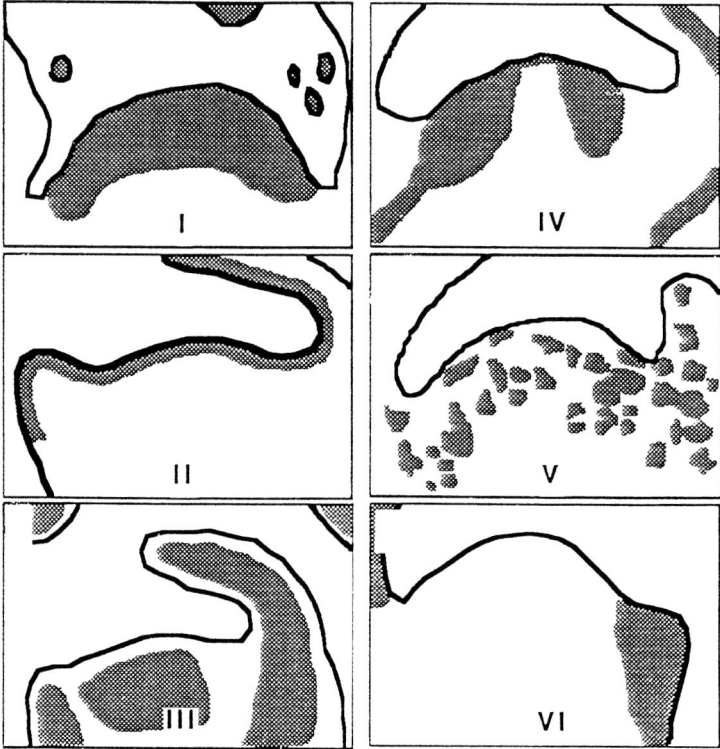

Figure 5.5 Patterns of gene expression in the tomato apical meristem. The patterns, and the genes showing different expression patterns, are listed in Table 5.2. From Fleming et al. (1993).

expression patterns of further genes in the shoot apex are listed by Medford (1992).

All these genes so far shown to be expressed in the shoot apex contribute to the picture that the shoot apex is heterogeneous and has different superimposed patterns of cellular structure, biochemical structure, and gene expression, as shown also by staining, distribution of macromolecules and enzyme activities. Differential expression of genes linked to the cell cycle would be expected to reflect differences in the cell cycle throughout the apex. The ground is being laid to explore the link between expression of specific genes and the morphogenetic activity of the apex, as in the case of the *knotted*, and related, genes. Perhaps attention should also be directed to genes specifying components of the cytoskeleton, and the synthesis of, and sensitivity to, plant growth substances in the apex. There may be very few genes with their expression restricted to the shoot

Table 5.2. *Expression patterns of genes in the shoot apex.*

Pattern of expression	Genes showing expression pattern
I. In the whole meristem	Ribosomal protein L2 (*rpl2*)
	Vegetative MADS-box gene
	Basic glucanase
	Shikimate kinase
	EPSP synthase
	18S rRNA
	Tm5
	Tm6
	S1
II. Higher in the tunica	Lipid transfer protein
	Polyphenoloxidase
III. Higher in the corpus	Arginine decarboxylase
	dUTPase
IV. Higher in the peripheral zone	Ribosomal protein L38 (*rpl38*)
	ypt2-like GTP binding protein
V. Higher in cell clusters	Histone 2A
	Histone 4
VI. In primordia but not in apical meristem	Small subunit of Rubisco

Data from Fleming *et al.* (1993).

meristem; these would be expected to be those genes conferring determination for shoot rather than any other sort of meristem.

Other genes now being identified are those that alter development and functioning of the apex. These include genes necessary for the formation of the apex at all (Barton and Poethig 1993), and others that alter the development of the apex and leaf initiation, and some that cause fasciation in the apex (Leyser and Furner 1992, Medford et al. 1992).

5.5. Shoot apical meristem organization

To summarize: the cytohistological zonation depends mainly on RNA distribution and this is reflected in the distribution of ribosomes, but this seems to be only loosely correlated with growth rate in the apical dome. If the zonation underlies growth rate, then the link between these and the cellular pattern will presumably be wall extensibility as determined by the structure of the cell walls. The cellular ultrastructure seems to be general, providing the housekeeping tools for the cells, but to become modified during cellular differentiation and on flowering. The only ultrastructural change that can be linked to apical functioning in the vegetative apex is the change in starch occurrence (see Chapter 6) and this reflects metabolic, rather than ultrastructural, differences. Although the key processes in apical functioning seem to reside at the metabolic and molecular levels, the distributions of metabolites, enzyme activities and gene expression have thrown up no consistent picture of how these are linked to apical functioning. The most promising line of inquiry would seem to be to search for genes whose expression correlates with particular features of apical functioning, such as the initiation of primordia (e.g. the *knotted* gene), or the changes in size of the central zone, or changes in the occurrence of starch. This may then provide a starting point for trying to unravel the metabolic processes that determine apical functioning and structure.

The Mechanism of Primordium Initiation

6.1. The formation of a primordium: how is a primordium recognized and defined?

To some extent our ideas about what is happening in primordium initiation are conditioned by how we recognize a primordium when it first forms. The formation of a primordium involves part of the apical surface, on the flanks of the apical dome, bulging upwards and outwards. The bulge is first apparent only as a slight irregularity on the otherwise smoothly curving surface of the apical dome, as seen under a dissecting microscope with reflected light, or as a slight bulge when seen with the scanning electron microscope (SEM) (Fig. 6.1). Whether a very small primordium is present is clearly a subjective assessment, but so long as this is always done by the same observer, it may none the less be consistent. This assessment can be made a little more objective in longitudinal sections by drawing a line isolating that part of the apical dome which departs from sphericity (Lyndon 1968a).

The primordium can be unequivocally recognized only when the leaf axil is formed. Hussey (1971a) developed a very sensitive way of measuring the presence of a primordium and its subsequent growth. He defined the axillary distance as being that between the incipient leaf (primordium)

117

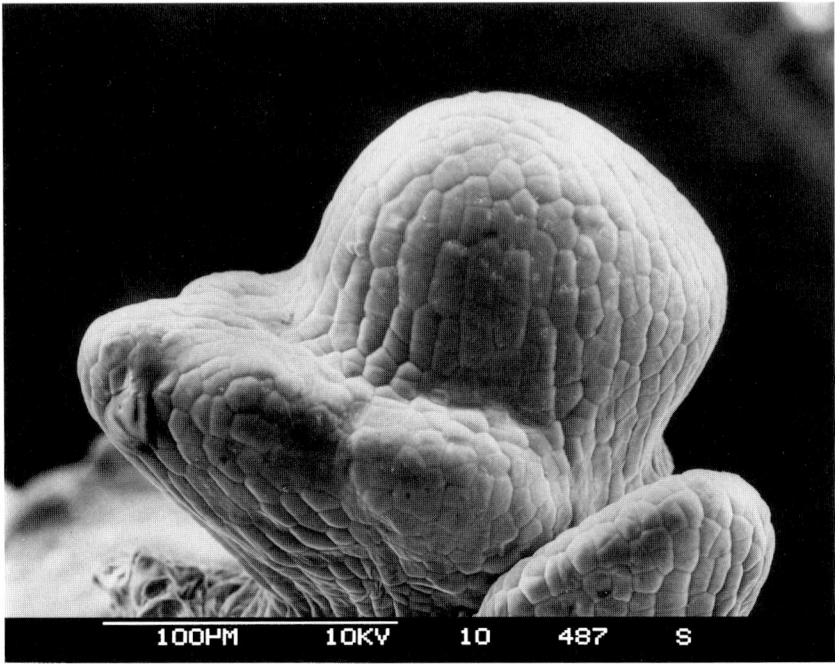

Figure 6.1 Pea shoot apex as seen in the scanning electron microscope. A new primordium is forming as a bulge on the right flank of the apical dome. The next oldest primordium (left) is beginning to form leaflets. Older primordia and leaves have been removed to reveal the apex. Bar $= 100\mu$m.

axil and a line drawn tangential to the tip of the apical summit and the tip of the newly forming primordium (Fig. 6.2). On this criterion, the primordium is formed once the axillary distance can be measured as a positive value, i.e. as soon as there is a measurable leaf axil. The progress of primordium growth can then be measured by the increase in the axillary distance, which tends to increase linearly with time. Although a potentially useful and sensitive measurement of plastochron stage, axillary distance has hardly ever been used. This may be because it is not well known, and because it requires either sections of the apex or exposure of the apex so that it can be measured under a dissecting microscope, and because it is most useful only over the first plastochron of a primordium's life. The definition of the stage of a primordium by the axillary distance focuses attention on the formation of the leaf axil, by growth being slower there than on each side of it (Hussey 1971b), so that growth in the adjacent tissues is seen as a corresponding bulge.

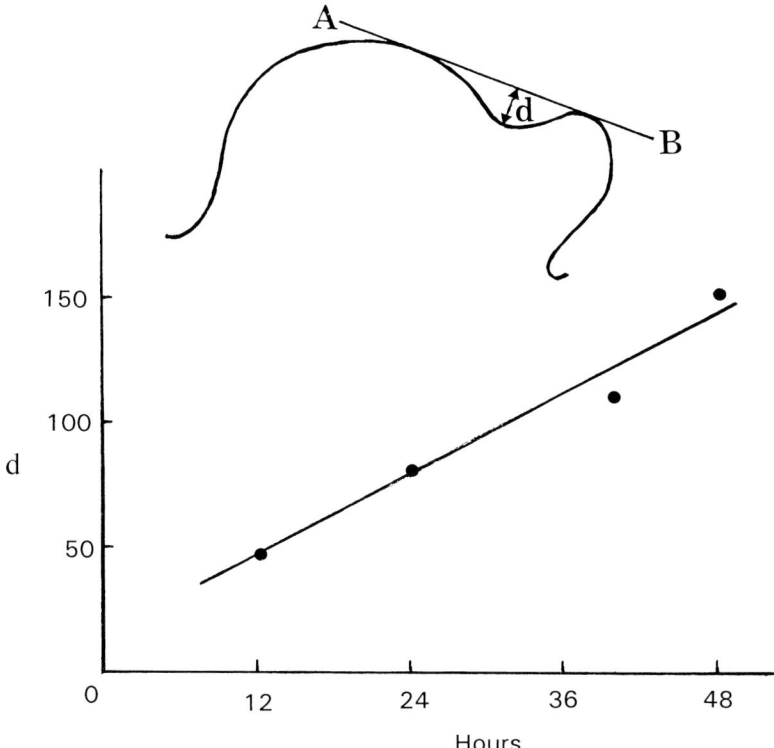

Figure 6.2 The progress of the plastochron can be measured from the axillary distance (d), which is the distance between the axil of the primordium and a line drawn joining the tip of the primordium and the surface of the apical dome. The axillary distance measured in tomato apices increases linearly with time through a plastochron. Data from Hussey (1971a).

The classical sign of primordium initiation is the appearance of periclinal divisions in the second tunica layer in angiosperms (or first layer in grasses). Such divisions are seen as soon as the apex has started to bulge and so confirm the presence of a primordium, although periclinal divisions can also occur in those parts of the apex which do not form primordia. The occurrence of periclinal divisions itself is therefore not sufficient evidence for a primordium forming.

Where the epidermis bulges up to form a primordium, so long as mean cell size remains constant and the leaf axils are fixed positions, then the rate of cell division in the epidermis must be greater at the bulging site than elsewhere on the apical surface, and this has been shown quantitatively for the pea (Lyndon 1976, 1982). In *Vinca* and

Anagallis, cell division was almost entirely at the primordium sites as seen from successive replicas of the apical surface of the apex as it continued growing (Green, Havelange and Bernier 1991, Williams 1991). Primordium formation can therefore be linked to an increased epidermal growth and cell division rate at the primordium site, although this does not necessarily mean that these events are causal; they could perhaps follow from turgor-driven expansion resulting from increases in surface plasticity.

A further criterion might be the change in arrangement of the surface cells or some characteristic pattern of divisions that could be seen with the SEM. The *de novo* formation of leaf primordia on watercress axillary explants, in the absence of a shoot apex, was preceded or accompanied by groups of T-divisions, i.e. new cell walls parallel to the files of cells rather than transverse to them (Selker and Lyndon 1996). Similarly, T-divisions have also been found at leaf sites in *Anacharis* (Green 1986) and *Silene* (Lyndon and Cunninghame 1986). In the *Hedera* apex, most of the cell divisions are transverse to the longitudinal cell files but those at the bulging sites of emerging primordia are commonly orientated longitudin-ally, i.e. T-divisions, or obliquely (Marc and Hackett 1992). However, in the *Vinca* apex the first divisions at the leaf sites were transverse to the cell files and divisions parallel to the longitudinal cell files (i.e. T-divisions) occurred only when the primordium had started to grow upwards (Williams 1991). The conclusion reached here was that 'cell shape, and subsequent cell divisions are governed by the major axis of extension growth' (Williams 1991, p.545). This may also be true of the other cases, since the change in division plane seems to coincide with the initial stages of the primordial bulging. The only example of a distinctive pat-tern of cell division appearing at a leaf site before associated bulging is on undetermined organ primordia in *Nasturtium* axillary explants (Selker and Lyndon 1996) but it has yet to be clearly shown how these develop into leaf primordia. This may be a special case of a cell division pattern preceding leaf formation, but which may not be typical of leaf initiation in general.

The formation of a primordium may also be detected by examination of epidermal wall structure with polarized light to show the mean orien-tation of the wall cellulose microfibrils. Primordium formation in *Graptopetalum* is then seen to be correlated with changes in the orienta-tions of microfibrils, and that where these become arranged in circles they prescribe the positions and extent of the new primordia (Green and Brooks 1978). Similar circle-forming by microtubules to delimit the primordium sites has been described in *Hedera* (Marc and Hackett 1989). Although primordium formation is therefore associated with microtubular and microfibrillar orientations in the epidermis, it is not clear whether these are causal to primordium initiation or simply accompanying events of a change in shape and surface stress that is brought about in other ways. That changes in microtubular and microfibrillar orientations anticipate, rather than accompany, organogenesis has not been shown. In sunflower, these changes follow, but do not precede, tissue undulation (Hernández and Green 1993).

Another criterion of whether a primordium is present comes from the work of Snow and Snow (1933). The primordium was regarded as determined about half a plastochron before a bump could be seen because about half-way through I_1 was when the position of the primordium could no longer be altered by making an incision in the apex. This implies that the causes of primordium formation must be sought when the determination of the cells becomes altered, before the bulging begins.

According to how the primordium is first recognized, the beginning and end of a plastochron may be defined differently by different workers. This does not matter so long as the stages in different work can be equated and it is recognized that the activity of the apical meristem is cyclic. What is clear is that there is no single criterion for when a primordium forms: however, the criteria used should always be clearly stated to allow comparison with other work.

6.2. Component processes of primordium formation

There are several processes that could be involved in primordium formation. The primordium could form because of (1) a locally increased growth rate, (2) a change in cell division plane and growth direction, (3) changes in the microfibrillar orientations of the outer epidermal walls that act as a restraining layer, (4) local changes in cell wall plasticity, and (5) a spatially periodic buckling of a uniformly growing tissue, or two or more of these processes could be acting in concert. Can some processes or events be identified that are more important than others? Indeed, are there some events or processes that are essential and others that are not?

Primordia could possibly be formed in a number of ways. The main possibilities are either a local faster growth rate, so that something like a pimple forms a 'growth centre' on the apical surface, or a change in the direction of growth, with local growth rate being unimportant. This in turn could be brought about in several ways:

1. The inner tissues could perhaps change growth direction to grow outwards at specific locations, forcing the surface to expand.
2. Alternatively, there might be generation of excess surface (an increased growth rate localized to the surface layers) so that the inner tissues are dragged out, being stimulated to grow out into the potential protuberance.
3. The yielding properties of the cell walls could change and be increased in some locations so that, in response to turgor pressure, the surface would tend to bulge out more in some places than others.
4. Conversely, the plasticity of the walls could become limited at the positions of the axils and at the summit of the apical dome and those parts of the surface where primordia do not form, so that primordia would be allowed to grow out only where the yielding properties of the wall were sufficient to allow it.
5. General, non-localized, greater growth or greater plasticity of an annulus round the apical dome could cause buckling of the surface and so result in local outwardly directed growth.
6. A local change in the plane of cell division could perhaps determine the major axis of cell extension so that the direction of growth overall is altered: periclinal divisions may be the determining factor in primordium formation.

We see that the primordium could be formed as a result of changes primarily in the epidermis, with the inner tissues responding, or by growth of the inner tissues, the epidermis growing passively to accommodate the underlying growth, and the changes could be because of changes in the rates of growth or the directions of growth.

6.2.1. *Increased growth rate*

The site of primordium formation is sometimes said to be a 'growth centre', but this is a vague term, implying a site of faster growth but begging the question of what is really meant, and being based on an assumption that the formation of a bulge must necessarily be the result of faster growth, but without any real evidence.

Table 6.1. *Rates of cell division in the epidermis and underlying cells of the median and lateral regions of the apical dome of the* Pisum *shoot apex.*

| | | Rate of cell division (% cells per h) | | | |
| | | I_1 | | Primordium | |
		A	B	C	D
Median 30μm	Epidermis	2.3	2.1	3.6	1.4
	Underlying cells	3.2	2.8	3.6	2.0
Lateral regions	Epidermis	2.5	2.4	2.9	2.1
	Underlying cells	2.4	2.6	2.9	2.1

Rates of division in the cells at the position at which a primordium will subsequently arise (I_1) and in the primordium itself, as they pass through two plastochrons, the first plastochron (A + B) (the first 30h [A] and the subsequent 16h [B]), and the next plastochron (C + D) (the first 30h [C] and the subsequent 16h [D]) (Lyndon 1982).

What then is the evidence for an increased growth rate at the primordium site? Direct measurements, by using colchicine, of rates of cell division have shown that the growth rate at the site of a new primordium was greater than anywhere else in the apex, in *Trifolium* (Denne 1966a), *Pisum* (Lyndon 1970a, Hussey 1972), *Lycoperiscon* (Hussey 1971b) and *Silene* (Lyndon and Cunninghame 1986), so there is clear evidence for a 'growth centre'. In the pea, however, more detailed measurements showed that the changes in growth rate are in fact not sufficient in themselves to account for primordium outgrowth. The rate of cell division in the incipient primordium was insufficient to account for the rate of increase in cell number in the primordium (Lyndon 1970a). The inevitable conclusion was that growth in other parts of the apex, especially the axial parts, contributed to the growth of the young primordium essentially by bulging out into it. In other words, in the pea the growth of the inner tissues was directed into the outgrowing primordium. Indeed, in the pea, periclinal divisions appear in the apex before primordium emergence, suggesting that a re-direction of growth is of major importance (Lyndon 1970b). An increased growth rate does, however, contribute to the outgrowth of the primordium. In the *Pisum* apex the rate of growth and cell division at the primordium site, especially in its midline, is greatest at the time of its emergence (Table 6.1; Lyndon 1982).

A greater growth rate at primordium sites implies a lower growth rate in the adjacent tissues. The formation of the primordium is correlated with the restriction of growth at the position of the axil, which eventually forms a crease between the primordium and the apical dome (Green,

Havelange and Bernier 1991). Williams (1975) questioned whether the axil was a fixed position. He suggested it was dynamic, i.e. a structure which has cells flowing through it, because the youngest primordium in tobacco can be at a lower level on the apical dome than the previous primordium. He believed this to be because the position of the axil moved upwards, as the young leaf increased in size faster than the adjacent apical dome tissue. If, on the other hand, the axil had been in a fixed position this would have implied that the apical summit became 'tipped over' to one side, moving from side to side, as in the pea apex (Nougarède and Rondet 1973). No direct evidence has been produced for tobacco, but marker spots placed on the axillary position on tomato apices were not displaced for at least one plastochron (Hussey 1971b). In the absence of evidence to the contrary, it must be assumed that, from its origin, the axil consists of cells which have differentiated there and remain there.

In the tomato apex, the epidermis at the site of an incipient primordium is under tension, as shown by the gaping of incisions made in the surface under carefully controlled conditions, and a group of cells with a higher division rate than elsewhere in the apex can be demonstrated just under the primordium site (Hussey 1971b, 1973). This led to the conclusion that in the tomato the primordial bulge was forced out by growth from within, being the result of the formation of a 'growth centre'. In the pea apex there is also a region of faster division just below the site of the incipient primordium (Lyndon 1970a; see Figs 3.4 and 4.1). However, the pea apical surface did not gape when cut; on the contrary, the cuts closed up, suggesting the tissues were under compression (Hussey 1972, 1973). These observations are consistent with the rate of cell division in the tomato tunica (the two outer cell layers) being two-thirds that of the corpus (Hussey 1971b) whereas in the pea the rates were the same in epidermis and underlying cells (Lyndon 1971). The data for the pea therefore do not support the idea of a primordium forming as a result of pressure from a 'growth centre', but could be consistent with the surface growth being faster than that required (i.e. two-thirds of the inner rate) to maintain a hemispherical shape, and so causing buckling of the surface because of 'excess' growth, as suggested by Green (1994).

6.2.2.   *Periclinal divisions*

When the occurrence of a primordium is scored as the appearance of a visible bump on the apical surface (Lyndon 1968a), then the principal cause of primordium formation appears to be a change in the direction of growth of the apical tissues at this point. The axis of growth is usually normal (at right angles) to the plane of cell division (Sinnott 1960, Lyndon 1990). The formation of a primordium is associated with the occurrence of periclinal divisions, which are normal to the axis of out-growth of the young primordium. The correspondence of periclinal div-isions with the site of primordium initiation is particularly clear in maize: the point at which a primordium arises is first seen as periclinal divisions in the epidermis. These then spread around the apex as the primordium grows and encircles the apex, the periclinal divisions marking exactly the lateral extent of the primordium (Sharman 1942).

But are the periclinal divisions cause or consequence? The view that 'the divisions initiating a leaf primordium cause the formation of a lateral protuberance on the side of the shoot apex' (Esau 1965, p.105) seems to imply that the process of primordium formation is understood and that periclinal divisions are the causative agents, but this is at least misleading, and probably wrong. This is because divisions do not cause growth directly. As Green (1976) has pointed out, divisions partition the cellular material created by growth. Divisions without growth simply cause the cells to become smaller. It is easy to see that in a non-growing system the plane of division would determine only the shape of the daughter cells, but this is also true in a growing system. The occurrence of periclinal divisions does not, and cannot, cause outward growth. When cell division is inhibited, the outgrowth of the leaf primordium can still begin (Foard 1971), as can the outward growth of the pericyclic cells in the initial stages of lateral root formation (Foard, Haber and Fishman 1965, Charlton 1977).

How closely are periclinal divisions correlated in time and in space with the initiation of a leaf? Usually observations have been restricted to median longitudinal sections and the occurrence of periclinal divisions in the tunica is seen to coincide with the formation of the foliar buttress. However, closer examination shows a more complex picture. In pea, periclinal divisions in the tunica are seen at the primordium site (I_1) before the apical dome begins to change shape at primordium emergence (Lyndon 1970b). In *Silene*, periclinal divisions can be found at all stages of the plastochron, and not only at the primordium site but also elsewhere in the apex, even where primordia are not formed, and so they cannot be used as a definitive indication of primordium formation (Cunninghame and Lyndon 1986, Lyndon and Cunninghame 1986). In the potato apex, periclinal divisions in the second tunica layer occurred not only at the primordium position but also all round the apical dome, not just in the regions of leaf initiation. Also, periclinal divisions were first evident in the corpus and indeed in the pith; periclinal divisions in the second tunica layer were often not seen until the primordium had already formed a small mound of tissue (Sussex 1955). Similarly, in the initiation of a phyllode in *Acacia*, the periclinal divisions associated with the primordial bulge are most obvious first in the corpus, and may often not be apparent in the tunica until the primordium has begun to grow out (Boke 1940). However, the observation that periclinal divisions are often seen first in the corpus, rather than the tunica (Lyndon 1970b, see references in Lyndon 1976), would in itself not preclude their being causal to primordium formation, simply that this would be a deep-seated rather than a surface phenomenon.

The distribution of periclinal divisions is therefore not always closely related to the position at which the primordium arises. The key question is whether the plane of division determines the direction of growth or whether it is a consequence of the growth direction. The arguments in support of the latter have been summarized by Lyndon (1990). Also, if periclinal divisions are not seen, this implies that all the divisions are anticlinal and the problem is not so much why periclinal divisions appear at the primordium site but why they have been suppressed previously (Lyndon 1972b). This poses the question of whether the formation of the leaf is an unusual growth phenomenon at a particular point on the apex, or whether it is because bulging on the flanks of the apex is the norm and the non-occurrence of primordia at other points on the flanks of the apical dome and on its summit are instead the result of restriction of growth at these points. At the position of the future leaf axil, growth slows down (Hussey 1971b). What is clear is that it cannot be assumed that because periclinal divisions are the first obvious indications of primordium formation, and change in growth direction, to the plant anatomist, they are therefore the causative agents or any more important than many other things that may be happening.

In the pea, periclinal divisions were absent in the I_1 region for the first half of the plastochron. Only anticlinal divisions occurred and these were equally in the plane of the median longitudinal section and normal to it (Fig. 6.3). In the second half of the plastochron, periclinal divisions occurred and their frequency was one-half that of the anticlinal divisions so that divisions were distributed equally in the three planes of space. Since this distribution was maintained throughout the plastochron, it was consistent with the division plane being random (Lyndon 1972b). The

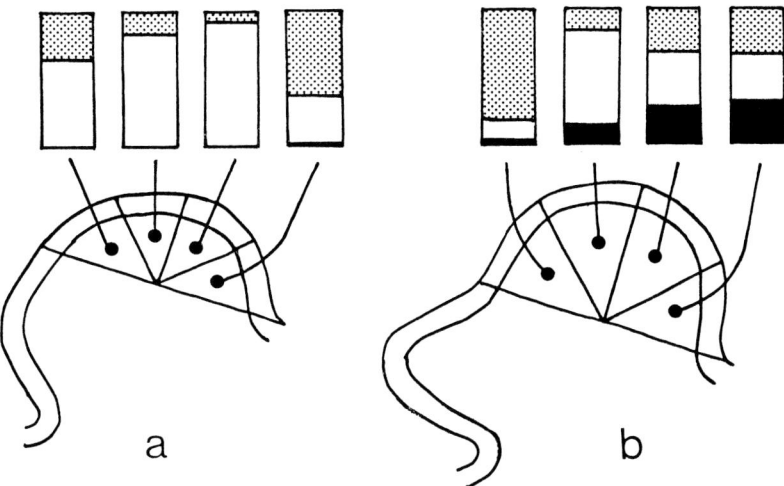

Figure 6.3 Periclinal divisions appear in the *Pisum* apex in the second half of the plastochron. In the pea apex (L.S.), divisions in the epidermis of the apical dome are all anticlinal and are not included in this analysis. In the first half of the plastochron (a) the primordium (left) is beginning to grow out. Divisions are overwhelmingly anticlinal perpendicular to the plane of the section (stippled) or in the plane of the section (white). In the second half of the plastochron (b), the primordium has grown further and periclinal divisions (black) have appeared throughout the apical dome but particularly in the I_1 position (right) where the next primordium is about to form. At this position, divisions are randomly orientated, since there are equal numbers of divisions in each of the three planes of space. Data from Lyndon (1970b).

problem is not, therefore, what promotes periclinal divisions, but what prevents them occurring in the first half of the plastochron in the pea? The formation of a primordium becomes, in fact, a problem of what prevents outgrowth of the apical surface before the primordium forms.

6.2.3. Changes in surface microstructure

Another possible mechanism for primordium formation involves changes in the outer epidermal wall microstructure so that its yielding properties become anisotropic and differ locally across the apex surface. The contention is that the surface microstructure determines the occurrence and positions of buckling of the surface so that primordium formation is essentially the result of the changing mechanical properties of the apex surface (Green 1980, 1994, Green and Poethig 1982). A refinement of this

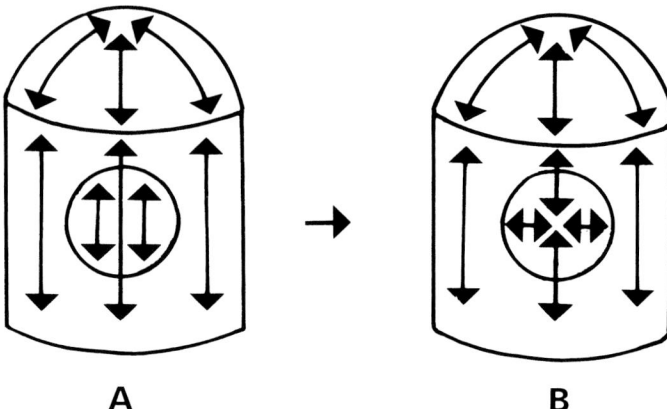

A   **B**

Figure 6.4 The formation of a new primordium on the flanks of the apex (A), essentially at right angles to the mother axis, requires that the predominantly longitudinal direction of growth associated with hoop-reinforcement, is modified through 90° at the sides of the new axis (B) so that it can also become radially symmetrical and hoop-reinforced. From Green and Brooks (1978).

point of view is that the surface is at first uniform in structure but that excess surface is produced by growth, and because it is constrained by the slower growing parts of the apex, the apical dome distal to it and the base of the apex proximal to it, it buckles, with a periodicity indicated by the formation of primordia and depending on the (uniform) mechanical properties of the buckling surface (Green, Steele and Rennich 1996). The anisotropy in surface structure would then follow, and correspond with, the change in shape of the surface.

The elongation of plant axes is principally in the axial direction and this appears to be the result of a restriction of lateral growth by the wall structure of the component cells. Because cellulose is birefringent, the average orientation of the cellulose microfibrils of the cell walls can be detected by polarized light and the dark and light regions of the preparation can be equated with specific orientations of the microfibrils (see Fig. 6.6). Microfibrils are found to be predominantly transverse to the shoot axis. The cell walls are thus hoop-reinforced so that lateral cell extension is limited and the principal direction of cell elongation is axial (Green 1980, 1984, 1994, Green and Poethig 1982). When a primordium forms on the flanks of the shoot apex, a new axis is formed essentially at right angles to the parent axis, so that the hoop reinforcement of the new axis (the primordium) must also become orientated at right angles to the parent axis. For this to happen, the main change in orientation of the wall microfibrils has to be at the sides of the new organ (Fig. 6.4).

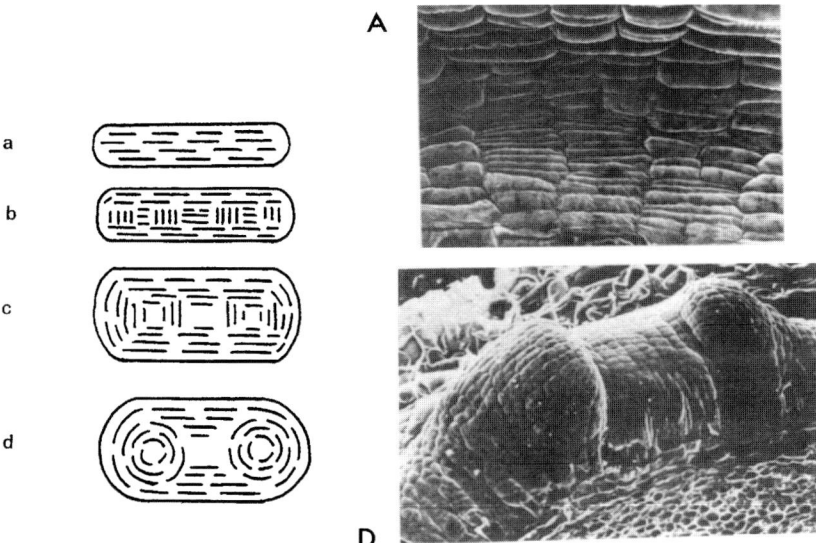

Figure 6.5 The formation of leaves and a new apex on the residual axillary meristem in *Graptopetalum*. The meristem consists at first of elongated cells (A) that show wall microfibrils predominantly transverse to the mother leaf long axis (a). At four positions on the meristem, microfibrils become reorientated through 90° (b), and then round off (c, d) to delineate the positions where the primordia form (d, D). From Green and Brooks (1978).

The detailed and elegant work of Green and his associates has shown that primordium initiation is indeed associated with changes in the orientation of the cellulose microfibrils in the outer surface of the epidermal cells. Primordia and a shoot apex are formed *de novo* on the residual axillary meristem of a *Graptopetalum* leaf when the leaf is detached from the plant. The mean microfibril orientation of the meristem surface cells is transverse (relative to the leaf axis) (Fig. 6.5a,A). Then at four positions the microfibril orientation changes through 90° so that two 'square' sites become delimited. At these sites the microfibrils then round off leaving a strip of longitudinally orientated microfibrils across the tip of the primordium site. Each primordium forms and then grows upward as a hoop-reinforced axis at right angles to the meristem surface (Fig. 6.5d,D), with a strip of cells, round the margin of the lamina, with microfibrils orientated transverse to the margin (see Section 6.4.1). The region of the meristem between the primordia becomes the shoot apex (Green and Brooks 1978).

In plants with opposite, decussate phyllotaxis, such as *Vinca*, the apex surface consists of two lateral lentoid areas, with microfibrils orientated

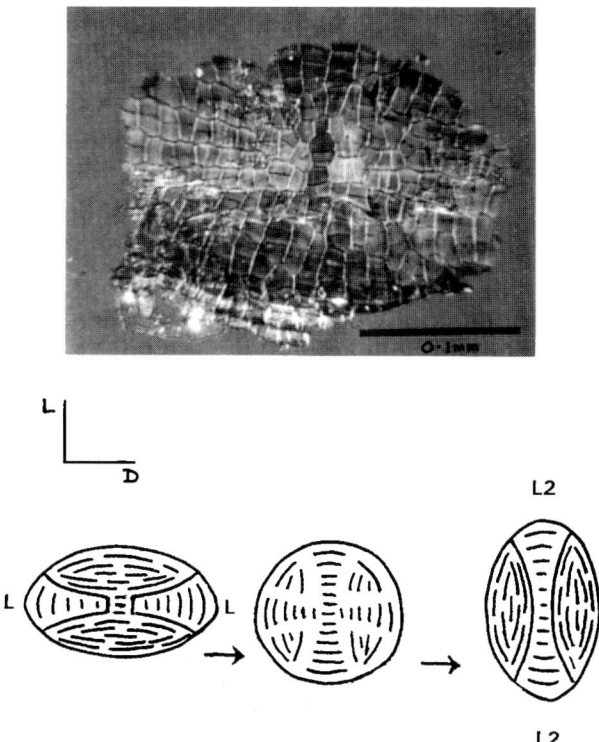

Figure 6.6 Microfibril orientations in the shoot apex of *Vinca*. The apical surface is carefully carved off the plant and viewed in polarized light so that microfibrils orientated predominantly north–south show up as light areas (L), and east–west as dark areas (D). The microfibril orientations in the apical surface layer are therefore as shown (left). A belt of cells with transverse microfibrils lies across the apex separating two lentoid areas with longitudinal microfibrils which are joined by a narrow corridor across the apical centre. The pair of new leaf primordia form at the ends of the surface ellipse (L, L). Then the apex undergoes reorganization, particularly at the ends of the lentoid regions, where the microfibrils reorientate through 90°, leading to the new apex, forming the next leaves (L2, L2), and at 90° to the former orientation. The process is then repeated. From Green (1985).

essentially along the length of each lentoid, separated by a corridor with a microfibril orientation transverse to the corridor and normal to the main microfibril direction in the lentoids (Fig. 6.6) (Green 1985, Jesuthasan and Green 1989). A new leaf pair forms when the microfibrils at each end of the corridor round off to delimit the primordium areas. The microfibrils at the tips of each lentoid and in the centre of the corridor become reorientated through 90° so that the whole structure becomes reconstituted but at 90° to the original. Repetition of these changes then leads to

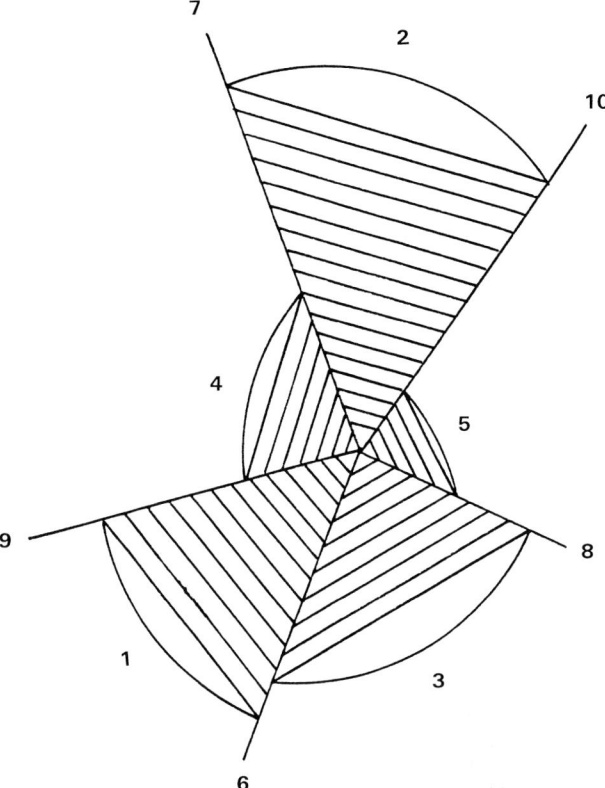

Figure 6.7 Orientation of microfibrils in the apical surface of *Ribes*, which has spirally arranged leaves. Each leaf and primordium (1–5) subtends a tangentially arranged microfibril field. New primordia (6–10) arise on the radii where two fields meet and there is therefore a discontinuity in microfibril orientation. The next primordium to be formed (6) arises on the radius where this angular discontinuity is sharp. From Green (1985).

the initiation of the next pair of primordia, at 90° to the previous pair (Fig. 6.6).

 Ribes is a plant with spiral (helical) phyllotaxis. Here the apex shows sectors, each subtended by a leaf or primordium and centred on the centre of the apex, and each showing tangential microfibril reinforcement (Fig. 6.7) (Green 1985). The sectors of tangential microfibril reinforcement seem to be those same sectors seen in surface views of the apex in which parallel cell files extend from the base of a developing leaf towards the apical dome surface, as seen in *Clethra* and *Ginkgo* (Hara 1971, 1980) and in *Hedera* (Marc and Hackett 1992). The radii for future leaves are marked by the discontinuities in alignment of microfibrils from one sector to its neighbours. The next primordium arises on the radius showing greatest angular discontinuity between adjacent sectors. What determines the position on the radius at which the primordium arises is not clear.

The formation of a primordium can therefore be envisaged as being caused by the buckling of the apical surface resulting from inequalities in surface growth and structure (Selker, Steucek and Green 1992, Green 1994). In flat or slightly domed apices the effective apical surface where primordia could form would be an annulus. In highly domed or elongate apices (as in grasses) the surface would approximate more to a cylinder. The wavelength of the buckling round the apical surface would determine how many primordia are formed at any one time (i.e. per plastochron). This in turn would depend on the properties of the surface: its microstructure (microfibril orientations) and its plasticity. The changes in reorientation of the microfibrils for the next cycle of primordium formation would be brought about by the tensions and stresses imposed on the surface by the lateral expansion of the adjacent older primordia and leaves and by the growth of the primordium and the apical dome itself.

The synthesis and orientation of the wall microfibrils almost certainly depends in some way on the underlying microtubules of the cellular cytoskeleton (Giddings and Staehelin 1991, Green and Selker 1991). Microtubule orientations parallel the orientations of the overlying microfibrils. It is the microtubules that are believed to be the strain-sensitive system in the cell. The orientation of the microtubules seems to depend on several specific cues: (1) direction of excessive tissue stretch, (2) orientation of division planes, and (3) orientation of microtubules in adjacent cells (Green and Selker 1991). The first and third cues are related to the directions and positions of existing growth and so may be determined ultimately by the plasticity of the surface and the orientation of other microfibrils. The dependence on division plane, though, implies that cell division itself may have an indirect role in determining microtubule and microfibril orientations and so the formation of a primordium. In subepidermal cells of *Graptopetalum*, a new polarity of microtubules was found only in cells that divided periclinally, normal to the surface and to the new axis of growth (Selker and Green 1984). It was not clear in this case what prompted the change in division orientation; it might have been a relaxation of pressure from the surface layers even though a visible outgrowth of these layers was not yet apparent. In the surface cells, changes in the orientation of the division plane may be the result of stresses imposed on the cells by their being stretched by the growth of leaf bases and developing primordia (Green and Selker 1991). This growth, in turn, may depend on the plasticity of the surface and on the presence of growth substances which can modify it.

It is still not clear whether the changes in the microfibril orientations on the apical surface, rather than changes in surface plasticity, are causal to primordium initiation rather than being consequences of changes in growth direction. What seems more certain is that microfibril orientations are related to the positions and shape of the resulting primordia. In *Graptopetalum* the primordia are essentially centric, whereas in *Pisum* they are dorsiventral and in *Silene* they arise as laterally spreading growths on the apical flanks. In each case the planes of division in the epidermis at the sites of primordium initiation are normal to the predominant growth axis (Lyndon and Cunninghame 1986). The orientations of the microfibrils and microtubules have not been examined in *Pisum* or *Silene*, but it might be expected that they would also correspond, as in *Graptopetalum*.

Further evidence that the orientation of microtubules and microfibrils is likely to be more concerned with precise positioning and primordium shape than with primordium initiation itself comes from the *fass* and *ton* mutants of *Arabidopsis* (Torres-Ruiz and Jürgens 1994, Traas et al. 1995). These mutants are unable to form orientated arrays of microtubules, which are instead randomly orientated throughout the plants. The plants themselves are stunted and barrel-shaped but, despite the lack of an orientated cytoskeleton and orientated divisions, they still form organs, including leaves and flowers, in their correct relative positions.

6.2.4. Change in plasticity of the surface

Basic to the initiation of primordia is the question of how the change in the direction of growth is limited to only certain sites on the apex where surface growth is greater than in the non-initiating sites. The converse is the lack of outgrowth at non-primordial sites. Although primordium formation, if it is the result of surface buckling, may not initially depend on local differences in surface plasticity, for continued growth the surface must be more extensible where primordia have formed than elsewhere on the apex. Cell wall growth is limited by its plastic extensibility, which it is known can be enhanced by auxin, especially in the epidermis (Kutschera and Briggs 1987, Kutschera, Bergfeld and Schopfer 1987, Masuda 1990,

Kutschera 1992). Plastic extensibility of walls has not been measured in the apex and it is difficult to see how it could technically be done.

Auxin and auxin analogues are, however, the main substances that have consistent effects in altering primordium size, which would result from local changes in surface plasticity. Morphactins act as inhibitors of auxin transport (Goldsmith 1977) and can also suppress primordium formation (Varkey and Nigam 1981). Application of 2,4-dichlorophenoxy-acetic acid (2,4-D) to the apex of *Phaseolus* stimulated formation of several primordia in place of the usual single primordium (Soma 1968). 2,4-D can also cause the formation of a circular collar round the apex, as though a primordium formed at all possible points around the flanks of the apex instead of at a specific location (Pereira and Dale 1982). This could be interpreted as the 2,4-D causing inhibition of basipetal auxin transport so that auxin accumulated at unnaturally high concentrations all round the apical dome and so caused increased plasticity of the epidermis all round the apex instead of only at the normal primordium site. N-1-naphthylphthalamic acid (NPA, a powerful auxin transport inhibitor) applied to an *Epilobium* primordium caused it to enlarge and so alter the phyllotaxis, whereas α-4-chlorophenoxyisobutyric acid (CPIB, an auxin antagonist) caused smaller primordia, again altering phyllotaxis (Meicenheimer 1981). Such alteration of primordium size may well have been the result of altered plasticity of the surface at the primordium site, NPA increasing it and CPIB decreasing it. This would be consistent with NPA preventing the basipetal movement of auxin out of the apex and so allowing it to accumulate and increase plasticity, and CPIB inhibiting auxin action and so decreasing the area over which plasticity was increased.

In some leafy liverworts, hydroxyproline causes formation of ventral leaves in species that normally lack them (Basile 1967, 1990). Hydroxyproline is known to block hydroxylation of proline residues in cell wall proteins which would prevent them acting as crosslinking agents in the cell wall. The walls would therefore become more extensible, so allowing primordium formation. Suppression of ethylene synthesis also allowed desuppression of leaf primordium formation but it is less clear what the mechanism might be (Basile and Basile 1983).

If the general plasticity of the primordium-generating surface (the annulus or cylinder round the apical meristem) depends on the action of chemicals (e.g. growth substances) on it, then it may be supposed that 'chemical activation determines whether primordia will form, physics determines where' (Green 1994, p.1786).

6.3. Is there a minimum cell number required for a primordium?

In mosses, where leaf formation appears to be a function of cell lineage (see Chapter 1), the number of cells involved in leaf initiation is presumably always the same, or very similar, for a particular species. In other plants, where the leaf and primordium can vary in size, is there a minimum number of cells required for primordium initiation?

Chimeras and clonal analysis can give information about (1) the numbers of cells involved in leaf initiation, and (2) the involvement of different layers and numbers of cells. In tobacco, mixtures of green, and mutant yellow or white, clones of cells were induced by irradiation of growing axillary buds (Poethig and Sussex 1985b). Different mutation patterns within the leaf showed that there was a minimum of three cells per leaf derived from each longitudinal file of cells passing over the leaf site on the shoot apex, and encompassing about 13 files in the transverse direction. This implies about 40 surface cells in the leaf primordium at initiation. Since there are also three to four cell layers involved (Poethig 1984) the total number of cells involved in initiation of a leaf primordium in tobacco is about 150. Similarly in maize, where the lateral extent of primordia is greater (and eventually encircles the apex), the primordium is about 45 cells wide, and is also about two to three cells in the longitudinal dimension so that about 100 surface cells are involved. With two to three cell layers also being involved, the total number of cells in an incipient maize leaf primordium is about 200 cells (Poethig 1984). Leaf primordium size at initiation in *Impatiens* has been estimated to be about 20–40 surface cells, and again, assuming three cell layers to be involved, this means about 100 cells per primordium (Battey and Lyndon 1988).

These values are comparable with some of those estimated from plastochron ratios (Table 6.2). Values derived from clonal analysis may sometimes be smaller than would be estimated by observing primordia once they have become visible (which is what plastochron ratio data provide) and perhaps indicate more accurately the number of cells involved at the time the leaf primordium becomes determined, which may be half a plastochron before it emerges (Snow and Snow 1933). By the time it becomes visible the primordium will already be growing. In potato, the formation of a leaf primordium was preceded by a lateral expansion of the apical flank where the primordium was about to emerge. As the leaf primordium enlarged it encroached 'laterally and distally over the apical surface, incorporating into itself tissues which were previously

Table 6.2. *Number of cells per primordium at initiation in vegetative shoot apices.*

Plant	Apex radius, r (μm)	Area of primordium relative to apical dome ($2 \log_e r$)	Primordial area on apical surface (μm^2)	No. of surface cells (at 100μm^2 per cell)	Total cells per primordium (assuming three cell layers)	
Epilobium	54	0.38	3000	30	90	
Xanthium	40	1.31	7000	70	210	100–
Impatiens	59	0.41	4700	47	141	200
Silene	59	0.71	7500	75	225	cells
Xanthium	87	0.95	23000	230	690	
Ranunculus	71	1.42	23200	232	696	c.700
Chrysanthemum	71	1.59	25000	250	750	cells

Calculated from data given in Lyndon and Battey (1985).

located at some distance from the centre of leaf inception' (Sussex 1955, p.265).

These estimates are of the minimal number of cells involved in a normal primordium. When primordia are smaller on initiation, as shown by a decrease in plastochron ratio and a reduction in absolute primordium area at initiation, presumably the minimal number of cells is also smaller, as it is in the initiation of the floral organs (see Table 8.2). There is also a minimal area required for a primordium to be initiated as shown by the isolation of small parts of the apical surface by vertical incisions (Snow and Snow 1952). In non-flowering *Lupinus* this was encompassed by an arc of 122° on the apical surface, but by a smaller arc in apices which were near to flowering, or in apices damaged by being split by a cut. This was consistent with the size of a primordium at initiation being determined, at least partly, by some influence from the apical summit which tended to increase the minimum area necessary for a primordium to form in the vegetative apex (Snow and Snow 1952, 1955). The minimum size for a primordium at initiation may therefore be expected to vary with the developmental stage of the apex and with the absolute size of the apex, as affected, for example, by nutrition of the plant and by temperature (Rogan and Smith 1975). Whether there is some ultimate lower limit, in terms of area or cell number, necessary for leaf initiation is not known, although it must be remembered that in mosses the primordium can be ultimately traceable to a single cell.

6.3.1. Contributions of different cell layers

It is well documented that the contribution of the various layers within the primordium to the final leaf structure can vary enormously but without affecting leaf form or the structure and differentiation of the leaf tissues (Tilney-Bassett 1986). The layer from which the cells originate seems irrelevant to their development (see Fig. 2.9). The only exception to this is the obvious one that the epidermis always originates from the outermost layer (L1), although L1 can sometimes also give rise to inner layers.

The epidermis (L1) may, however, exert an influence on the development of the inner layers. In *Pelargonium*, a one-layered skin can apparently destroy the function of the pistil or transform normal stamens into petaloid staminodes (Neilson-Jones 1969). In a *Camellia* chimera, the presence of the epidermis of *sasanqua* over a *japonica* interior causes or allows the formation of stamens and carpels, whereas if the epidermis and inner layers are all *japonica* no stamens or carpels are formed. The epidermis can also produce a unique result: in *Camellia* 'one new character found in the graft chimera, and not in either component or component species, was the presence of six to eight styles instead of the three typical of *C. japonica* and *C. sasanqua* species' (Stewart, Meyer and Dermen 1972, p.521).

In *Nicotiana* chimeras, the epidermis had the major influence on flower morphology (Marcotrigiano 1986) but, conversely, the internal cells (L3) determined floral meristem size and carpel number in tomato (Szymkowiak and Sussex 1992). These various examples all point to all layers of the apex contributing to primordium formation, but the genes controlling different aspects of primordium formation may be exerting their action through diffusible substances which allow one layer to influence development in adjacent layers. This is similar to what seems to happen in the development of Kranz anatomy in C4 leaves (Nelson and Langdale 1989).

6.4. Determination of the primordium and its parts

The cells comprising the primordium become determined, i.e. their developmental fate as primordial cells becomes unalterable, sometime during early primordium development. Snow and Snow (1933) found that by making incisions in the *Lupinus* shoot apex the position of the next primordium could be altered, but only if the incisions were made more than half a plastochron before the primordium emerged. The primordium was therefore considered to become determined half a plastochron before it began to form as a bulge.

The elegant experiments of Sachs (1969), who made incisions in young pea leaf primordia, showed that the parts of the primordium become gradually determined as the primordium develops. Partial ablation or mutilation of the primordium when it was a barely noticeable bulge on the apex did not prevent the primordia developing normally; at this stage they were capable of complete regeneration. Primordia that were 30μm long were still capable of partial regeneration, although the positions of the leaflets and tendrils and stipules had by this time become fixed. But once they were 70μm long, about one plastochron after initiation, the primordia were no longer capable of regenerating excised parts. The pea primordium therefore seems to take until about one plastochron after its initiation to become fully determined. Determination is therefore a gradual and hierarchical process, the parts of the primordium becoming determined after determination of the primordium as a leaf.

Primordia of the fern, *Osmunda*, were excised and cultured (Steeves 1966). The older they were when excised, the less likely they were to develop as shoots. When they were about nine plastochrons old they seemed to be completely determined as leaves. To check whether, when a shoot was formed, it was perhaps developing from axillary cells of the primordium rather than by transformation of an undetermined primordium, Cutter (1954) marked the apical cell of excised leaf primordia of *Dryopteris* with Indian ink. Whether the primordia developed as leaves or shoots, the apical cell was always marked and so must have been the original primordial apical cell. The primordia had therefore transformed directly into either leaves or shoots and could not have been determined at the time of excision.

6.4.1.   *Origin of polarity and dorsiventrality*

In pea, the predominantly longitudinal polarity of the epidermis seems to become imposed on the underlying cells as the primordium grows (Lyndon 1982). The growth rate on the abaxial surface of the primordium may also be slightly greater than on the adaxial surface, as shown by the greater frequency of cell divisions on the abaxial surface in pea (Lyndon 1982), so that the primordium grows upwards. In many plants

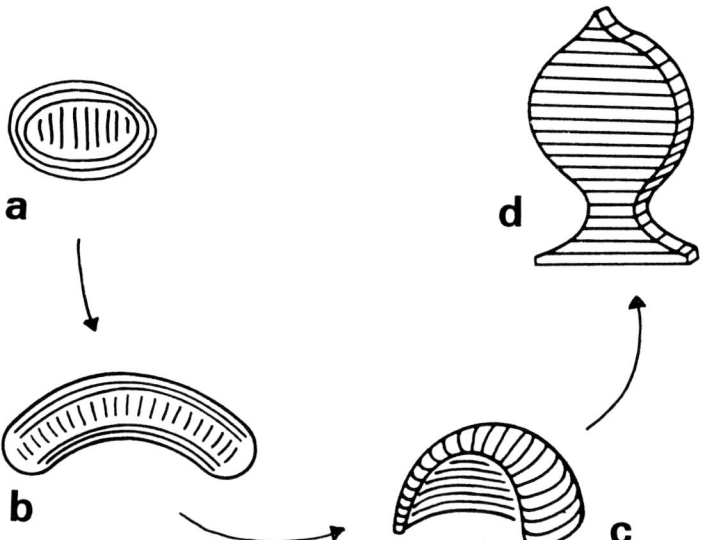

Figure 6.8 Leaf formation typically involves the formation of a belt of cells across the leaf site having a transverse microfibril orientation (a). As the primordium widens (b) and grows upwards and outwards (c), these cells give rise to the margin of the leaf (d). From Green (1986).

the primordia grow upwards to enclose the apex. In the tomato, the young primordium grows upwards as the area of the apical surface that contributes to the abaxial surface is greater than that contributing to the adaxial surface, as shown by the displacement of points on the apical surface, so that even if growth rates are the same on both sides of the primordium it will still tend to curl upwards (Hussey 1971b).

Dorsiventrality may become established at the time the primordium is initiated. The initiation of a primordium on the apical surface involves the reorientation of microfibrils at the primordium site. The centre of the primordium site is crossed by a corridor of cells, tangential to the apical centre, with microfibrils orientated radially to the centre of the apex (Fig. 6.8) (Green 1986). This corridor separates the upper and lower surfaces of the primordium and eventually forms the cells of the margin around the edge of the leaf. These cells, with markedly longitudinal orientation, can be seen clearly in some mature leaves, such as tobacco (Poethig and Sussex 1985a) and *Silene* (Fig. 6.9).

Related observations are those of chimeras with chlorophyll-deficient cells which formed leaves with clones of longitudinal sectors, which did not necessarily correspond on the abaxial and adaxial surfaces of the leaves (Dulieu 1968, 1969). Also, the characteristics of the upper and

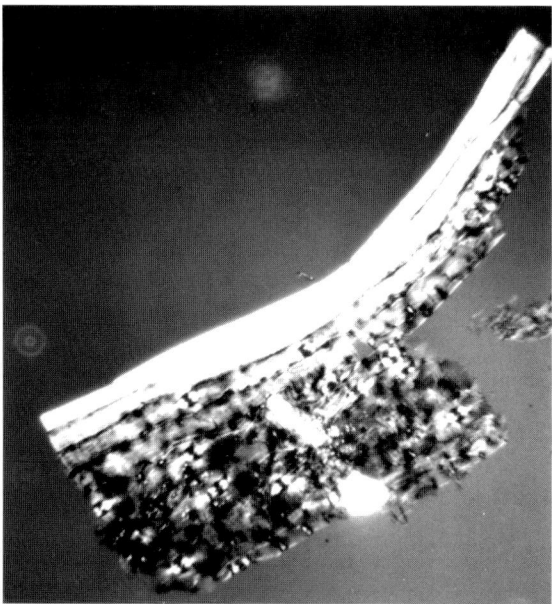

Figure 6.9 The leaf margin of a *Silene* leaf viewed in polarized light. The cells of the leaf margin are elongated and have microfibrils orientated transversely to the cells' long axis.

lower surfaces of the leaf can be determined by different genes. A group of genes, *wlo*, *wb* and *wsp*, limit the distribution of wax to either the upper or lower surface of the leaflets in *Pisum* (Marx 1977). This is all consistent with the upper and lower surfaces of the leaf originating from different groups of cells in the primordium, indicating that dorsiventrality is probably established when the primordium is initiated.

6.5. Control of primordium formation

An increased growth rate at the primordium sites cannot be essential for leaf initiation since primordia are initiated perfectly well in young pteridophyte and in floral apices without a growth gradient and where the growth rate on the flanks of the apex is essentially the same as at the summit (see Chapter 3). The higher growth rate on the flanks of the meristem, together with periclinal divisions there, will allow the outward bulging of the apical surface (Lyndon and Cunninghame 1986) associated with the establishment of wall microstructure which favours deformation of the surface there (Green 1986). But it is doubtful if these are other than enabling factors in primordium formation. The key factor is surely the ability to change the shape of the apical surface as a result of changes in the tensile properties or plasticity of the surface. Theoretically this could be achieved either by the action of a diffusion–reaction mechanism in which the primordia were sites of production of a wall-loosening agent, or by a physical mechanism in which strain re-orientation of the wall microfibrils caused potential weak spots on the surface which would preferentially buckle outwards under turgor pressure. Alternatively, the action of a diffusion–reaction mechanism or a biophysical mechanism could be uniform over the whole potential primordium-generating area of the apex, the formation of primordia depending on the wavelength of the buckling of the surface. What seems likely is that chemical factors determine whether a primordium can arise, i.e. whether the epidermal walls at potential primordium sites have an increased plasticity which will allow local outward bulging, but that the biophysical factors determine where the primordium arises (Green 1994). The general position of primordia could still be determined by a diffusion–reaction mechanism which would 'anchor' the peaks of the buckling wavelength in relation to existing apical structure. The occurrence of periclinal divisions is probably a response to the changed growth direction at the primordium site, the cells perhaps dividing in the plane of least shear, normal to the new growth axis (Lintilhac 1984, Lyndon 1990).

7

Positioning the Primordia

7.1. Phyllotaxis: leaf positioning

A striking feature of shoots is very often the pattern of the leaf arrangement, or phyllotaxis (Jean 1994). It is determined by the arrangement of primordia at the shoot apex, except for those slight alterations in the leaf arrangement which result from accommodation to the underlying vascular structure and from subsequent uneven growth or twisting in the internodes. The typical vegetative phyllotactic patterns are whorled, spiral, opposite and decussate, and distichous. Where the leaf arrangement is orthogonal (i.e. opposite or distichous), and successive leaves are at 90 or 180° from each other, the leaves are positioned above older leaves in vertical ranks or orthostichies. Descriptions of such arrangements can be precise. Even when there are whorls of leaves, the number per whorl and whether successive whorls alternate, is usually sufficient to describe leaf arrangement accurately. The real problems arise in describing spiral leaf arrangements, in which the divergence angle between successive primordia often approximates to 137°, and the many primordia on an apex form a pattern of intersecting spirals, termed parastichies. In such patterns, orthostichies are absent, because no leaf primordium is formed directly above a previous leaf. The numbers of parastichies in each of the two intersecting sets often correspond to two successive numbers of the series 1, 1, 2, 3, 5, 8, 13, 21, . . ., etc., where each number is the sum of the two preceding numbers. This series of numbers was first described by the Italian mathematician Leonardo Fibonacci ($c.$1170–$c.$1250) in the thirteenth century and hence is often referred to as a Fibonacci sequence. When expressed as a series of fractions, 1/2, 2/3, 3/5, 5/8, . . ., etc., this infinite series approaches a limit which is $(\sqrt{5} - 1)/2$. The divergence angle between successive primordia, in an apex with parastichy numbers belonging to this series (and where the parastichies intersect orthogon-ally), is $360° \times [1 - (\sqrt{5} - 1)/2]$, i.e. 137.507 . . .°, the Fibonacci angle.

Until 1950, descriptions of phyllotaxis depended on which phyllotactic theory they were based. This was why Richards (1951) proposed a description of phyllotaxis which was objective and did not depend on any presupposed theory. He derived the Phyllotaxis Index which, together with the divergence angle, could define phyllotaxis at the shoot apex, including not only spiral but also orthogonal and multijugate systems. However, the usefulness of this has been questioned, since in many plants the parastichy spirals do not intersect orthogonally and the divergence angle may not be exactly the Fibonacci angle (Thornley 1975b). A more useful and general description is given by the divergence angle and the plastochron ratio which defines phyllotaxis completely, as pointed out by Richards (1951).

Over 30 theories of phyllotaxis were summarized by Schwabe (1984), and even more by Jean (1994), who produced a new theory of phyllo-taxis based on the principle of minimal entropy production. Included in this theory, which is expressed mathematically, is the notion that rhyth-mic behaviour is intrinsic to biological systems. This theory allows predictions of which phyllotactic systems can and do exist and provides a more complete description of phyllotactic patterns. It explains the basis for the patterns and the positioning of primordia but does not attempt to explain what causes the positioning of the primordia other than the positions of previous primordia. However, there are irregular and regular arrangements in *Magnolia* (Zagórska-Marek 1994) that do not conform to the mathematical theoretical expectations outlined by Jean (1994). These exceptions to the rule must ultimately be interpret-able to understand properly the mechanism of the generation of phyl-lotactic patterns.

In a developmental context, description of phyllotaxis is useful only in so far as it leads to testable hypotheses about the mechanism of primor-dium positioning. Jean (1994, p.144) believes that the problem of phyllotaxis is 'the problem of understanding the purpose served by the orderly arrangements of primordia'. However, this holistic and teleological approach is unlikely to prove helpful in understanding the processes involved in primordium formation and positioning. The basis for the origin of phyllotaxis must be looked for at lower hierarchical levels of organization than the pattern itself. The many different phyllotactic patterns have no obvious selective advantages and it has not been possible so far to discern any functional significance for them that cannot be over-ridden by compensatory morphological changes such as internode elongation (Niklas 1988), although Fibonacci phyllotaxis may be most effective for light interception in rosette plants and distichous and opposite phyllotaxis best when internodes develop. Phyllotaxis seems to exist because it is the inevitable outcome of processes concerned with primordium formation; what we need to understand is what these processes are. This seems to be much better addressed by descriptions of phyllotaxis such as those of Richards (1951), which lead to hypotheses about the mechanisms of primordium formation which can be tested. We still require an explanation which can eventually be traced back to metabolic reactions and the activity of the genes.

7.2.   Primordium positioning in the floral apex

When a meristem makes the transition from producing leaves to producing flowers, there is a change in the arrangement and positioning of primordia on the meristem (Lyndon 1998). Flowers characteristically have whorls of organs, with four or five (dicotyledons) or three or six (monocotyledons) members in each whorl. On transition to the flower meristem, the phyllotaxis (using the term to cover all types of apical organs) and divergence angle alters, except in those plants with whorled leaves (and even here the numbers per whorl may differ in vegetative growth and in the flower). Divergence angles (between successively arising primordia) in the flower are typically $120°$ in three- and six-membered whorls, 90 or $180°$ in four-membered whorls, and 72 or $144°$ in five-membered whorls. Since the phyllotaxis 'is uniquely defined by divergence angle and plastochrone ratio' (Richards 1951, p.516) a change in the divergence angle also implies a change in the phyllotaxis or primordium arrangement. If there is also a change in plastochron ratio then this in turn implies a change in the size of a primordium on initiation when this is taken to be the proportional increase in apical surface that is required for a new primordium to form (see Chapter 8).

The flower typically consists of successive sets of floral organs: sepals, petals, stamens and carpels. In the flower, most phyllotactic patterns are whorled, although many whorls may originate as a helix or spiral (Sattler 1973), and the angular displacement of one whorl from the previous one is usually half that of the angle between adjacent members of the older whorl. In some plants, such as *Echeveria* (Green 1988), the whorls are orthogonal because each consists of four members, although successive whorls alternate with each whorl being displaced 45° from the previous one. A very important difference between vegetative and floral phyllotaxis is that in the flower there are usually no internodes developed between successive primordia or whorls of primordia. This may be one of the factors determining differences between vegetative and floral phyllotaxis in a given plant, since the vertical distance between primordia at initiation on the apex may be an important determinant of primordium positioning. This is perhaps because increased vertical distance between primordia reduces the influence of the previous primordium on the formation of the next one (Schwabe 1998). In the *flo* and *squa* mutants of *Antirrhinum*, axillary buds with a spiral arrangement of bracts replace the flowers of the wild-type, but the positioning of the primordia in these axillary buds is almost identical to that of sepals (Carpenter et al. 1995). Presumably internodes develop in the buds of the mutants but not in the flowers of the wild-type.

In plants with a terminal flower, the phyllotaxis of the flower often seems to continue the phyllotaxis of the inflorescence or the leafy shoot. This is particularly striking in grasses where there can be a continuous phyllotactic sequence from the leaves to the carpel of the terminal spikelet (Sharman 1947). In other plants with a terminal flower the phyllotactic sequence may continue from the leaves into the flower, although the pattern may become modified, as has been detailed for *Silene*, in which the leaves are opposite and decussate but the floral organs are initiated helically, and mature as whorls (Lyndon 1978a, b).

In *Silene*, the axillary buds of each pair of leaves are unequal, and the larger ones of each pair form a helix up the stem (as do the smaller ones). In the first stages of flower formation the apical dome enlarges (Miller and Lyndon 1976) and the large axillary buds just below the apex also enlarge (Lyndon 1978a). The first pair of primordia of the flower is smaller than leaf primordia and the members are not quite opposite each other but displaced towards the adjacent axillary bud and separated

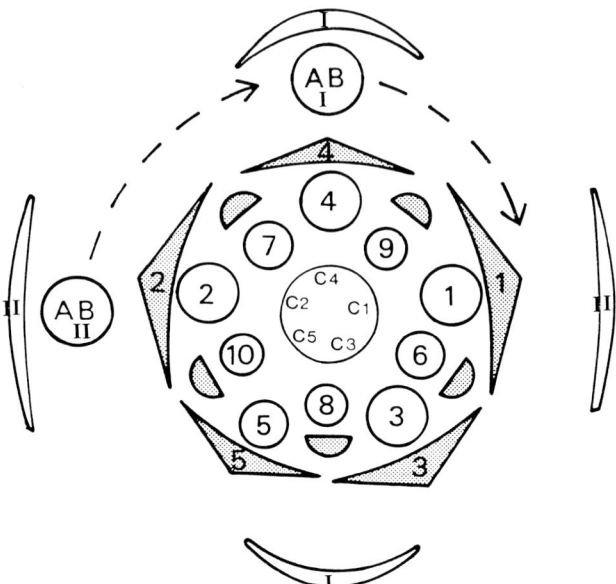

Figure 7.1 Positioning of primordia in the flower of *Silene coeli-rosa*. The first sepal (1) forms at the position where the next bud would be expected in the upward spiral of large, unequal axillary buds (ABII, ABI). Sepal 2 is formed across the apex, but at 156°, not at 180°, as are the leaf pairs (II, II; I, I). Sepal 3 forms across the apex but nearer the older sepal, so establishing the direction of the spiral of the floral organs, which is opposite in direction to the spiral of axillary buds. The petals form on radii between the sepal positions and the stamens form on the same radii as the sepals or the petals, but appear in spiral sequence (1–10). From Lyndon (1978a).

by 156°; they become the first two sepals (Fig. 7.1) (Lyndon 1978b). The sepal that is positioned so that it is a direct continuation of the helix of the larger axillary buds is itself larger. The third sepal is initiated at a position which is nearer this larger sepal and so establishes the direction of the spiral for the initiation of the remaining sepals and the rest of the floral organs. This spiral is always opposite in direction to the upwards helix of large axillary buds. This means that the positions of all the floral organs can be precisely identified and allows comparison of the same floral organ from flower to flower. This is why only in *Silene* has it been possible to follow the development of individual floral organs and in a way which has so far not been possible for any other flower (Figs 7.2, 7.3) (Lyndon 1978a, b, 1979a).

The stamens are the next primordia to become visible in *Silene*, although all the petals become visible about when the first three stamens are formed (Lyndon 1978a). The first five stamens are initiated in a spiral, with a mean divergence of 144.9°, which results in a whorl of five in which adjacent primordia are, on average, 72° apart. The second whorl of stamens is also initiated in a continuation of the spiral, but with the sixth stamen only 103.9° from the fifth (close to $72 + 36°$, i.e. $108°$) so that the next five stamens form a whorl which alternates with the first five stamens. Despite being initiated in a helix or spiral, the petals, stamens and carpels all originated at positions that were best defined as being on the radii which bisected the angle between the primordia of the adjacent outer whorl of organs (Lyndon 1978b). The formation and early development of the *Arabidopsis* flower has also been documented in detail (Smyth, Bowman and Meyerowitz 1990). The photographs of flower development documented by Sattler (1973) suggest that the developmental patterns of the *Silene* and *Arabidopsis* flowers are far from unique and are probably representative of many plants. Many other patterns of flower development, differing in detail from the general scheme, may also be found (Leins, Tucker and Endress 1988).

7.3. Origin of phyllotaxis during ontogeny

The positioning of primordia is usually considered in the context of adding further primordia to a pre-existing pattern. But it must be asked how positioning is determined when primordium formation first starts in the absence of an existing pattern. The leaf arrangement characteristic of a species is self-generating, as shown during micropropagation by the ability of new shoot apices, arising on callus in culture, to form with the normal leaf arrangement in the absence of a pre-existing phyllotactic pattern.

The commonest situation in which phyllotactic patterns form apparently *de novo* (free of the influence of existing pattern) is when a bud forms a flower in a racemose inflorescence where bracts do not form below each flower, e.g. as in *Arabidopsis*. Each floral meristem then forms whorls of floral organs in the phyllotaxis characteristic of the flowers of that species. Although the first whorl (sepals) forms in the absence of previous primordia on that meristem, the use of positional information is implied by the flower being characteristically in a defined orientation with respect to the subtending axis. This is acknowledged in floral diagrams when the position of the axis is shown relative to the flower structure and pattern.

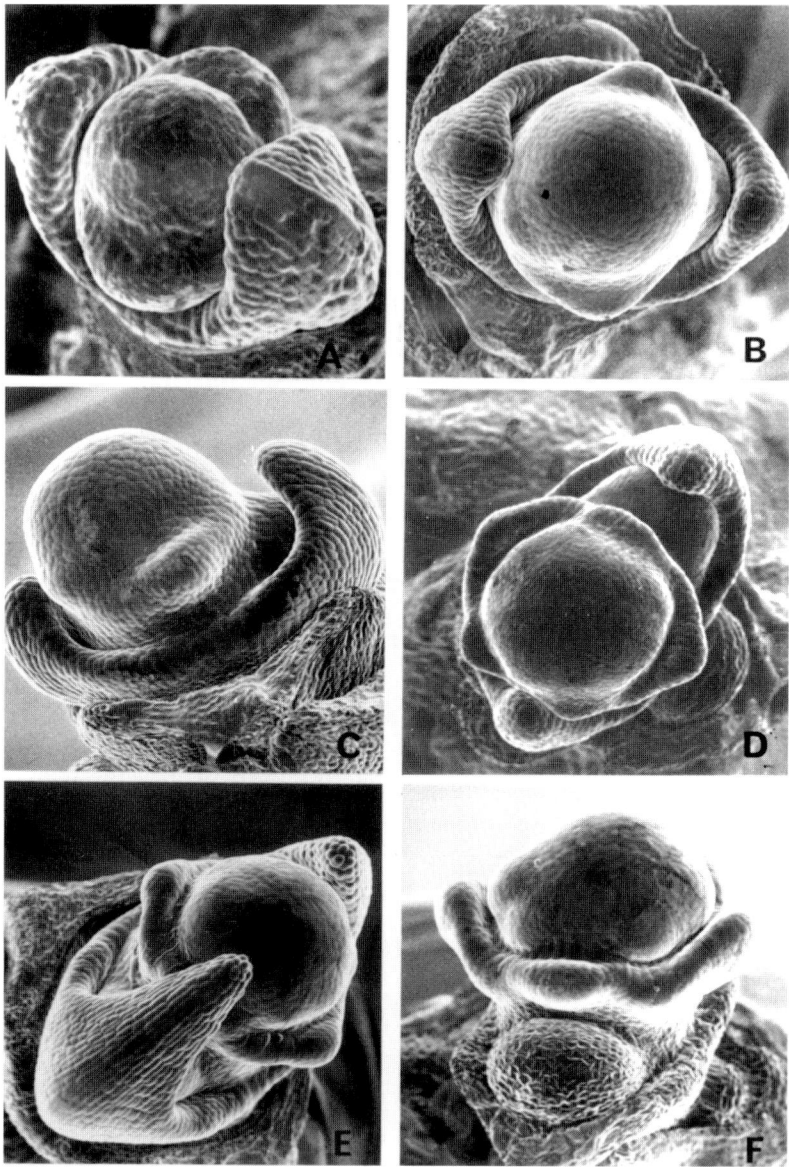

Figure 7.2 Early development of the flower of *Silene coeli-rosa* as seen in the scanning electron microscope. (A) In the vegetative apex the leaf primordia are initiated in opposite pairs, so the divergence angle between leaves is 180°. Each new pair of leaves is at 90° from the previous pair. (B) The first two sepals are not quite opposite but at 156° from each other; the first sepal is slightly larger than the second. The third sepal is just forming to the left of the first sepal, and slightly above it on the apical dome. The last axillary bud is to the right of the apical dome in the leaf axil. The spiral of buds is clockwise in this apex, and the

There are few examples of leaf primordia being formed *de novo* on a bare apex. In the dicotyledonous embryo the first leaf primordia form on the bare apex above the cotyledons. In many plants with spiral phyllotaxis this usually starts as decussate, with the first leaves being at 90° to the cotyledons, which therefore apparently provide positional reference points. Only during later development does the phyllotaxis make the transition to spiral (as can be clearly seen in plants such as *Helianthus tuberosus*). In axillary shoots, the axil itself provides positional information as shown by the observation that the pattern of leaf initiation in the bud is characteristically orientated with respect to the axil. A key question is how the cotyledons arise as an opposite pair in the embryo in dicotyledons. The chirality of subsequent phyllotaxis usually seems to be random (e.g. Rijven 1968, 1969). The handedness of the flower in *Silene* was always opposite to that of the helix of large axillary buds (Lyndon 1978a), but equal numbers of plants were right- or left-handed.

One situation where leaf primordia do arise on a bare apex with no obvious positional information on the apex is in meristems in the axils of watercress leaves (Ballade 1970). In each leaf axil there are many meristems, one of which normally develops into the axillary bud and the rest develop into roots. However, if the axillary surface with its primordia is excised and cultured, some of these primordia will develop as shoots instead of roots. They then produce leaves in a spiral, as in the normal shoot apex, but in the absence of any preformed shoot pattern, which must arise during the development of the young meristem.

In watercress axillary explants a new bud may also form *de novo* at the front of the explant on a small bare area, especially in explants cultured in the presence of cytokinin (Selker and Lyndon 1996). The positions at which primordia formed, and the phyllotactic pattern which they

Caption to Figure 7.2 *(cont.)*
sepal spiral anticlockwise. (C) The same apex as (B) viewed from the side. The first sepal is pointing to the observer, and the third sepal is now clearly visible at the left of the base of the apical dome. (D) All five sepals have formed. In this apex the sepal spiral is clockwise. The last axillary bud is larger, and this will carry on the growth of the inflorescence, which is a scorpioid cyme. The stamens are just beginning to form. (E) The sepals have fused laterally and will eventually form the calyx tube. (F) Just above and alternating with the fused sepals, the petals are forming as pendulous bulges. The fourth sepal is towards the observer, and just above the last axillary bud. The first few stamens are forming, and stamen 1 is evident next to sepal 1 at the left. From Lyndon (1978a).

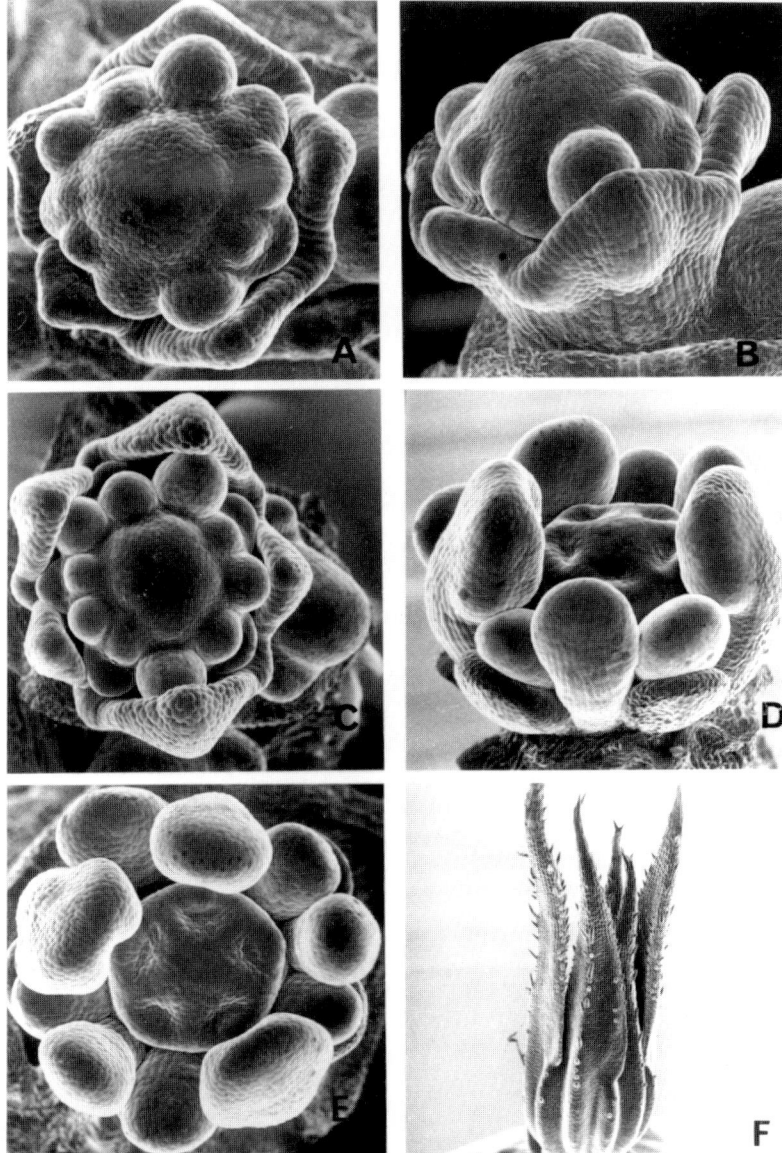

Figure 7.3 Later development of the flower of *Silene coeli-rosa* as seen in the scanning electron microscope. (A) The first five stamens (opposite the sepals) have all been formed. The spiral in this apex is clockwise. Of the next five stamens, which alternate with the sepals, 6 and 7 are evident, 8 and 9 are just discernible as small bumps and 10 has yet to be initiated. Just below the positions of sepals 6–10 the petals are present as primordia alternating with the sepals. (B) The same apex seen from the front. Stamen 2 is towards the observer, but stamen 10 has not yet been initiated above the petal which is

assumed, seemed to depend on the positions of creases in the explant surface, and therefore seemed probably to be regulated by the mechanical stresses in the surface at the leaf-generating sites. But what regulates the initial size of the leaf primordia or the number of cells involved is not known. These leaf primordia were formed in the absence of an organized shoot meristem; this seemed to develop subsequently and did not seem to be necessary for leaf initiation (Selker and Lyndon 1996). This would be consistent with the shoot meristem being principally a continued source of cells for the growth of the shoot, and also providing a platform where organogenesis can occur.

7.4.   Localization of primordium formation

The possibility of a minimum number of cells being required for primordium formation was discussed earlier (Chapter 6), but what regulates the upper limits of primordium size? Why is a primordium localized to just a part of the apical surface? Why does a new primordium not form as a ring or an annulus all round the apex? Primordia are not formed at all on the apical summit until carpel formation, and normally do not arise all round the apex, but are limited in size at initiation and form only at predictable positions on the flanks of the apex. In special cases, a continuous primordium does indeed form all round the apex, as in *Phaseolus* apices treated with 2,4-D (Pereira and Dale 1982). However, complete whorls of fused primordia in flowers usually form initially as individual

Caption to Figure 7.3 *(cont.)*
immediately to the left of stamen 2 and is protruding between sepals 1 and 5. (C) All the sepals, petals and stamens have been initiated (clockwise spiral) and the apex has become approximately pentagonal, just before carpel initiation. The different sizes of the sepals and stamens are a result of their being initiated in sequence but with similar growth rates, so that the size differentials are maintained throughout much of flower development. (D) Apex with sepals removed. The stamens have grown faster than the petals, two of which are seen here at the base of the flower. The carpels have been initiated as indentations on the apical surface. (E) Top view of the same apex as (D): the spiral in this apex is counterclockwise. The margins of the carpels will grow up and fuse laterally and the partitions between the carpels will not develop, so giving rise to the free-central placentation of the ovary. (F) A later flower bud showing the different sizes of the sepals, which are maintained through to the mature flower. The same inequality of the mature sepals can be seen in roses, for example. From Lyndon (1978a).

primordia which subsequently fuse laterally (Sattler 1973). Another unusual case is that of a double mutant of *Antirrhinum* (*flo/squa*), where very occasionally shoots are formed in which the two primordium sites produce primordia continuously so that two sets of continuous, helical bracts wind up the stem (Carpenter et al. 1995). This is reminiscent of Plantefol's theory that leaves were produced by two generative centres moving round the apex, but for which there was no clear evidence (Cutter 1959).

Primordium size at initiation could perhaps be determined by the minimum number of cells required. Once the apical dome has generated this number of cells sufficiently far from the apical summit, then the primordium would form. If this is all that were required then a whorl of leaves could form, i.e. a primordium all round the apex but split up into discrete bits. If, however, the primordia are formed singly, this implies that it is not simply distance from the apical summit that determines whether a primordium can form, although a minimum distance may be one of the enabling requirements. There must be some other requirement that is fulfilled at one position on the apical surface but not at another. This is seen especially when primordia are initiated on alternate sides of the apex as in a distichous arrangement, or at precise positions at divergence angles one from the other of approximately 137° as in the characteristic Fibonacci arrangement. Since phyllotaxis is orderly and patterned, the important requirement for determining the position of a new primordium is presumably the positions of the previously formed primordia on the apical surface. The problem then is how existing primordia determine the position of a new primordium.

There have been many models proposed to account for leaf positioning. They fall into two main categories: (1) packing theories and (2) field theories. The packing theories propose that there is mechanical interference between existing leaves and primordia on the apical surface and the sites of new primordia. Primordia would therefore occupy sharply defined areas. The field theories propose that the site of a new primordium is determined by the concentrations there of one or more inhibitors or promoters of leaf initiation, these being produced by the existing leaves and primordia, and/or by the shoot apical dome. The area of a primordium at initiation would therefore be defined in relation to the distance of its centre from the centres of adjacent primordia.

7.4.1. Packing theories

Observation of the shoot apex showed that a new primordium formed only on that part of the apex which was free from primordia, and that the primordia do not impinge upon each other but each develops a regular geometrical form. The primordia appeared and grew as though each had its own defined space in which to grow. It seemed that as soon as the apex had grown sufficiently then a new primordium was formed in the 'first available space' (Snow and Snow 1931, 1933). This hypothesis assumes that the primordium at initiation has some specific surface area and that until this area is available on the apical surface a primordium cannot arise. Richards (1948) pointed out that this should preclude the over-lapping of primordia or congenital fusion of primordia, which had been observed as a result of surgical operations on the apex and aberrant plants in the field. Any geometric theory of primordium positioning eventually requires a way of defining the primordium size in cellular and physiological terms. This may now be possible in terms of the structure of the epidermal surface cell walls, as shown by Green (see below).

7.4.2. Diffusion–reaction mechanisms

The simplest of the field theories derives from Richards' (1951) analysis of phyllotaxis, from which it can be concluded that the sites of new primordia arise at sites that are:

1. a minimal distance from the summit of the apical dome,
2. as far away as possible from the previous primordium, and
3. distant from the neighbouring primordia in the inverse ratio of their ages in plastochrons.

To account for a constant phyllotaxis it is also necessary to assume that

4. the primordium size at initiation is constant, and
5. the rate of radial expansion of the apex is exponential.

All of these conditions are consistent with what is observed at the shoot apex. Conditions 2 and 3 suggest that the primordia are mutually inhibitory and this led to the proposition that the newly formed primordia were sources of an inhibitor of further primordium initiation and that the effects of this inhibitor decreased as a function of distance from the sources of production.

Models based on this premise – that the newly formed primordium is a source of inhibitor of initiation of further primordia, and that the effect of the inhibitor diminishes with distance – produce typical phyllotactic patterns (e.g. Thornley 1975a, Veen and Lindenmayer 1977). These models assume that the flanks of the apex, where the primordia are formed correspond to an essentially cylindrical surface and so avoid the question of why primordia do not form toward the summit of the apex. This is addressed by models that postulate the same or another inhibitor being produced by the apical dome, as in the model of Chapman and Perry (1987), which proposes that primordium formation results from the interaction of such an apically produced inhibitor with a promoter diffusing upwards from the vascular system. Models in which an interacting promoter and inhibitor are both produced by new primordia give remarkable simulations of alternate and opposite phyllotaxis (Meinhardt 1982), and can be applied to simulating spiral phyllotaxis by assuming additional lateral interactions between primordia (Mitchison 1977).

A remarkable purely physical model has been devised to simulate phyllotaxis (Douady and Couder 1992). Drops of iron colloid suspension are allowed to fall regularly onto an upturned point in the centre of a pool of oil on which the drops float and remain discrete. The whole system is placed in a vertical magnetic field so that the ferrofluid drops act as magnetic dipoles and mutually repel each other. As the drops move outwards and also away from each other under the influence of the magnetic field, more drops are added one by one at the centre. New drops slide down the central point and move off in a direction that depends on the positions and magnetic influence of the pre-existing drops. What is remarkable is that the device generates a pattern of outwardly moving drops which form parastichies just as in Fibonacci phyllotaxis, and the pattern depends on the interval between the addition of drops (the 'plastochron') in relation to the rate of outward radial movement. With a long interval between successive drops a distichous pattern appears and if the interval between drops is reduced the pattern becomes one of parastichies, with more complex 'phyllotactic' patterns the more rapidly drops are added. Opposite decussate patterns can also be generated (Douady and Couder 1992). What this model reveals is that the phyllotactic patterns observed are consistent with new primordia behaving like mutually repulsive drops.

In the shoot apex, the interaction between primordia seems unlikely to be because each is a magnetic dipole. What physiological mechanisms could result in the primordia behaving in this way?

All models involving the production, by a newly initiated primordium, of an inhibitor of further primordium initiation make the assumption that sites of inhibitor production are formed *de novo* at minima in the inhibitor concentration field (e.g. Thornley 1975a). This recalls habituation in callus, where autonomous production of a growth substance can be initiated by reducing the concentration of that same substance in the tissues (Meins and Lutz 1980, Meins 1983). If this is indeed what is happening in the shoot apex, it may be indicative of a much wider role for habituation-like processes in plant development (Jackson and Lyndon 1990). No such inhibitors have yet been identified in shoot apices, although there are pointers to their existence and to what they might be (see Section 7.7).

7.5. Biophysical mechanisms: changes in surface microstructure

Changes in the orientation of the cellulose microfibril arrays in the outer epidermal cell walls of the shoot apex are found at the sites of new primordia. Is the position of a primordium determined by the microfibrillar pattern and, if so, how is this itself determined?

7.5.1. Vegetative apex

In *Graptopetalum*, leaf primordia form at two sites on the residual meristem, first delimited by two pairs of sites at which wall microfibril orientation changes from transverse to longitudinal (see Fig. 6.5) (Green and Brooks 1978). Each pair of sites is where a primordium forms but what initially determines these sites at which microfibril orientation first changes, is not clear. They may arise as a consequence of the buckling of the surface, this being the result of an increased plasticity of the whole of the meristem surface, the sites and distance between them being determined by the general properties of the buckling surface (Green 1994). If this is so, then the primordium sites may be outlined by the microfibrils but they may be determined by the yielding properties of the apical surface.

In spiral phyllotaxis, as in *Ribes*, primordia arise on the radii formed by adjacent tangential microfibril fields (subtended by adjacent leaves and primordia) where they abut on each other at a characteristic angle. Where this angle is greatest marks the radius on which the next primordium will form (see Fig. 6.7) (Green 1985). It is not clear how the microfibrils then round off to form the leaf site nor what determines the radial distance at which this happens or how the size of a primordium at initiation is delimited. As the leaf bases grow laterally they are thought to stretch the surface of the apical dome so that the stress fields become realigned. Mechanical stresses imposed by growth of the older, subtending leaves are therefore thought to be crucial in determining surface buckling, and the subsequent changes in microfibrillar orientations, and therefore the positions of new primordia. This further suggests the importance of folds and creases in the meristem structure so that primordium positioning would depend on the precise patterns, rates and directions of growth of the young leaf bases (Green 1985, Selker and Lyndon 1996).

Position, however, could result from the characteristics of the system rather than by the definition of precise coordinates. A spontaneous physical pattern of crests and troughs in the surface layer would be expected to originate if more growth in an annulus round the flanks of the apex were constrained by less growth above and below it (the summit and base of the apical dome respectively). This model has been illustrated by a knitted strip in which periodic undulations appear which are the result of a greater mass of material between the centre and sides of the strip, although the strip is uniform along its length (Selker, Steucek and Green 1992). This has been modelled in detail for whorl production on an annulus (Green, Steele and Rennich 1996). Similarly, a metal plate compressed at its perimeter bulges non-uniformly (Chai 1990), reminiscent of primordium formation. When the shoot apex is non-hemispherical then the shear stress and tension is also non-uniform on the apical surface. In an ellipsoidal apex, as in plants such as *Vinca*, with opposite, decussate leaf arrangement (see Fig. 6.6), the new primordia are formed at sites of high shear stress and low tension (Selker, Steucek and Green 1992). It is not yet clear whether this also holds for other apices that form only one primordium at a time.

7.5.2.   *Flowering apex*

In some plants, such as the Magnoliaceae, the spiral pattern of leaf formation is continued into the flower. But in many flowers the primordium arrangement in the flower differs from that in the vegetative shoot. *Echeveria* is a succulent dicotyledon in which the leaves are arranged

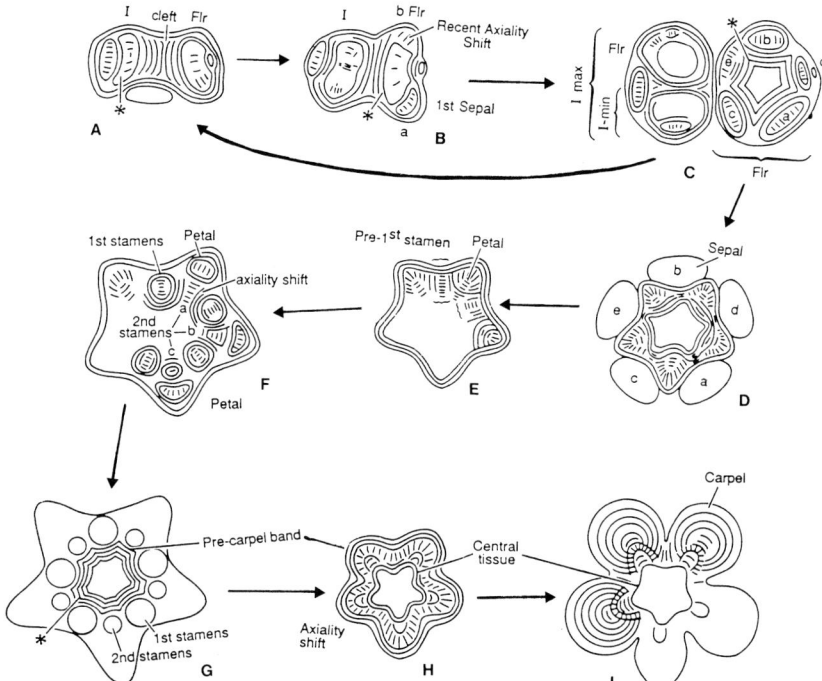

Figure 7.4 Orientation of microfibrils in the surface of the inflorescence (I) and flower (Flr) apex of *Echeveria*. The inflorescence meristem (I) forms a flower (Flr) and a new inflorescence meristem (A–C). The asterisks (*) indicate sites where axiality shifts will occur to give radial cellulose reinforcement. For further explanation, see the text. From Green (1988).

spirally, with a divergence of 137.5° (Green 1988) but as the inflorescence forms the bracts subtending the flowers are at 180 or 90° to each other, and in the flower the organs are arranged in successive whorls of five, so that the average divergence between organs in a whorl is 72°. The transitions between these different arrangements are thought to be triggered by changes in the stress patterns on the apex brought about by changes in the rate of lateral stretching of the subtending leaves and bracts (Green 1988).

The *Echeveria* flowers are formed singly, in succession, by the inflorescence meristem, which is roughly rectangular (Fig. 7.4A). It bifurcates transversely and the cleft that thus separates the continuing inflorescence meristem (I) from the new flower meristem (Flr) is characterized by microfibrils being orientated predominantly along the cleft (Fig. 7.4B) and cell divisions also orientated longitudinally. The inflorescence meristem then repeats this cycle at 90° as the floral meristem develops (Fig. 7C,A). Of the two bracts subtending the inflorescence meristem, one is large and the opposite one is small, but why there is this difference in size is unknown. The first sepal forms next to the large bract, and the first two sepals are displaced away from the midline of the flower and away from the subtended inflorescence meristem (Green 1988). (This is unlike *Silene*, in which the first two sepals are displaced *towards* the inflorescence meristem [Fig. 7.1] [Lyndon 1978a]. This may be because the flower meristem seems to grow faster than the inflorescence meristem in *Echeveria* [Green 1988], but slower in *Silene* [Lyndon and Cunninghame 1986].) The positions of the first sepals in *Echeveria*, as they form, are marked by an axiality shift in the microfibrils towards the outside corners of the flower meristem (Fig. 7.4B,C) and these are followed by axiality shifts at the positions of the subsequent sepals (Fig. 7.4C,a,b,c,d,e), which originate in spiral sequence but which rapidly become a more uniform whorl by more rapid growth of the later formed sepals so that all sepals soon become similar in size. (This is again unlike *Silene*, in which the unequal sizes of the sepals are conserved through into the mature flower [Lyndon 1978a].) The axiality shift at the sepal positions results in microfibrils at 90° to the prevailing hoop-reinforcement pattern. Similar axiality shifts occur in positions just interior to, and alternating with, the sepals, and these denote the petal positions (Fig. 7.4D,E). These shifts result in the sepals and petals having hoop-reinforcement but with radially orientated microfibrils over their tips (Fig. 7.4C,F) (Green 1988), in the same way as leaves do (see Fig. 6.10) (Green 1986).

Stamen formation in *Echeveria* shows microfibril axiality shifts, from circumferential to radial, at the positions *between* the potential stamen primordia, so that when the stamens form they have a band of transversely orientated microfibrils over their tips, unlike any other of the floral organs (Fig. 7.4F). It is not clear why the microfibrils in the case of the stamens round off to include the transversely orientated microfibrils at the centre of each organ rather than rounding off to include the radially orientated microfibrils as in all the other organs. This implies that there may be some other factor at work, rather than microfibril orientation alone, in determining the positions of the floral organs. However, the result is that the radial band of transversely orientated microfibrils over the tip of each stamen marks the furrow that develops between the lobes of the anther. The carpels form, like the sepals and petals, with accompanying axiality shifts from transverse to radial (Fig. 7.4G,H,J) (Green

1988). If the shifts in surface microfibril orientation are the result of stresses in the surface brought about by differential growth of the existing organs, it remains to be explained what determines and localizes these growth rates. They do, however, seem to depend on the previously established pattern, so that the positions of the primordia of one floral whorl seem to depend on the positions of the primordia of the previous whorl.

Similar changes in microfibril orientations occur during flower formation in *Kalanchoë* (Nelson 1990a, b) except that axiality shifts over the positions of the tips of the primordia were not observed until the primordia had begun to form. Over each sepal and petal primordium a narrow band of radially orientated microfibrils formed, but over the centre of the stamens, and the carpels, the microfibrils tended to remain unaligned. The stamens developed as centric structures (Nelson 1990b). The first formed (inner) stamens formed from common primordia each of which formed a stamen and a petal; the outer stamens formed from primordia each forming only a stamen. The microfibril alignments for petal/stamen primordia and for later stamen primordia were similar so that the same microfibril alignments seemed to be common to different types of primordia. The changes in microfibril orientation took place, as in *Echeveria*, at positions apparently dictated by the positions of the primordia of the previously formed whorls of primordia.

In *Kalanchoë*, sometimes flowers with an unusual symmetry occurred. Often this was because the vegetative phyllotaxis had been tricussate (leading to six-membered whorls in the flower) or spiral (leading to five-membered whorls). Abnormal arrangements and numbers of floral organs in some whorls but not others were produced by growing plants under conditions of light stress, so that the pattern of later whorls was not completely dependent on the pattern of the previous whorls (Nelson 1990b). These and the previous observations, together with similar abnormalities and meristic variation, that occur in many plants (Meyer 1966), and the occurrence of homeotic mutants (Coen 1991), indicate that, in general, the microfibril orientations associated with a particular type of primordium may (1) not be specific to that primordium type, and (2) may not depend completely on the positions and microfibril orientations of the preceding floral whorl. Since the changes in microfibril orientation seem to accompany rather than precede primordium formation (Nelson 1990b) it can be argued that (3) they may be accompaniments of primordium formation rather than its cause. It has also been argued that the microfibril orientations are more likely concerned with the shape of the resulting organs rather than determining the positions at which

Table 7.1. *Volume relative of growth rate of successive floral whorls, at initiation, in the* Silene *flower, for plants grown at different temperatures. Growth rate appears to alternate between whorls (Lyndon and Cunninghame 1986).*

Temp (°C)	Volume relative growth rate (per day)			
	Sepals	Petals	Stamens	Carpels
13	0.47	0.39	0.49	0.27
20	0.64	0.56	0.66	0.37
27	0.57	0.26	0.34	0.26
Temp (°C)	Volume relative growth (per day) relative to sepals = 1.0			
	Sepals	Petals	Stamens	Carpels
13	1.00	0.83	1.04	0.57
20	1.00	0.88	1.03	0.58
27	1.00	0.46	0.60	0.46

they are formed (Lyndon and Cunninghame 1986). Even if microfibril orientations are the result of stresses due to differential growth rates of the existing organs at the shoot apex, it still remains to be explained how the growth rates are controlled locally. For instance, the rate of growth of successive floral whorls in *Silene* apparently differed, in a characteristic manner (Table 7.1) (Lyndon and Cunninghame 1986) but the final form of the flower (and therefore the relative positions of the organ primordia) was the same irrespective of different growth rates and sizes of primordia at initiation in plants grown at different temperatures (Lyndon 1979a).

The arrangement of microfibrils on the apical surface therefore delimits the positions of the primordia, but it is not clear what reorientates the microfibrils at specific positions on the apex. If it is the stresses and tensions produced in the apical surface by the growth of the apex and the existing primordia, then it must be understood how this in turn is controlled in an orderly way so that precise phyllotactic patterns are generated. The ability of the *Arabidopsis* mutants *ton* and *fass* to make leaves and flowers with the primordia apparently in the expected pattern, although the orientation of microtubules and, presumably, microfibrils is random (Torres-Ruiz and Jürgens 1994, Traas et al. 1995), implies that the microfibril orientations are more likely to be an accessory of primordium formation and positioning rather than its prime cause, and to be concerned more with determining the shape and growth axis of the primordia as they form and develop.

7.6. Procambium and primordium positioning

The positions of primordia could also conceivably be determined by the positions of procambial strands which differentiate in the tissues below the shoot apex. Procambial strands in *Populus* can be detected below the primordium positions up to 14 or more plastochrons before the primordia are initiated (Larson 1983). However, because of the complexity of the vascular system in these *Populus* trees, it could be argued that it is difficult to find a position on the meristem that is not above a procambial strand. Procambium cells can be detected histochemically in *Pisum* and *Vicia* shoot apices just below and at the sites of primordia being initiated and can also be found at sites of primordia before there is any sign of primordium formation (Gahan and Bellani 1984). Incisions in the shoot apex of *Euphorbia* could alter the positions of the primordia and of the procambium (Soma 1958). When a primordium was displaced, the differentiation of the procambium to the new position could be seen before the primordium was formed. It was suggested that the procambium was therefore determining the position of the primordium. However, if the procambium was the first to be visibly differentiated, but the primordium was determined equally early but was not visible as a bump until half a plastochron or so later, then it may be that the same factors were responsible for the positioning of the primordium and the procambium independently. In *Equisetum*, leaf number can vary from whorl to whorl with corresponding changes in vascular structure, but it is not clear whether procambium position determines leaf position, or vice versa (Bierhorst 1959).

To see if leaf primordium positioning was affected by the procambium in *Lupinus*, horizontal cuts were made in shoot apices just below the I_2 or early I_1 positions, that is, before the positions of the primordia were fixed (Snow and Snow 1948). Although the procambium was disrupted, the primordia formed essentially in their normal positions, and procambium differentiated from the primordia downwards to the cut, but then stopped. The positions of the primordia were therefore apparently not determined by the procambium in the stem.

By making four vertical cuts, a shoot apical dome could be isolated on a plug of pith tissue (Ball 1952b). As the shoot apex regenerated, procambium was formed not only below the first primordium to form, but also in the apex above the level of insertion of the primordium, suggesting that the shoot axis itself, as well as primordia, can induce procambium. It seems probable that procambium is induced in both primordia and axis and that these procambial strands join by the sort of mutual attraction that also determines the way in which fascicular and interfascicular cambia link up at the beginning of secondary thickening.

7.7. Experimental modification of primordium positioning

Phyllotaxis depends on the plastochron ratio and the divergence angle (Richards 1951). Whenever there is a change in the size of the primordia relative to the apical dome (i.e. the plastochron ratio), then there is a change in phyllotaxis. In a few cases, a change in size of the apical dome or the primordia has been produced experimentally, resulting in changes in angular divergence of successive primordia.

The ABPHYL (aberrant phyllotaxis) mutant in maize has a greater number of leaves than a normal maize plant, and decussate or spiral phyllotaxis instead of distichous. This was correlated with narrower leaves, only two-thirds the width of normal leaves and originating from fewer cells on the circumference of the apical dome than in normal plants (Greyson and Walden 1972). The ABPHYL apex was also 17% wider than the normal (Greyson et al. 1978). The changed phyllotaxis was apparently due to a reduction in the size of primordia at initiation relative to the size of the apical dome.

Actual experimental attempts to test theories of phyllotaxis and the control of primordium positioning are much rarer than the theories. Probably all experiments so far have attempted to find support for a particular mechanism rather than to try to distinguish between different possible mechanisms. This was not true of the earlier surgical work, but even these experiments could not be designed to allow critical distinction between packing and field mechanisms.

7.7.1. *Chemical modification of phyllotaxis*

If diffusion–reaction mechanisms are reponsible for patterns of leaf initiation then it should be possible to affect them by upsetting the

Table 7.2. *Effects of TIBA on apical growth in* Chrysanthemum *(Schwabe 1971)*.

Plants about 6–11 weeks after treatment with TIBA	Untreated	TIBA-treated
Plastochron (days)	4.8	8.1
Mean divergence angle (between three youngest primordia)	137°	180°
Apical angle	64°3′	44°45′
Mean volume RGR (per day)	0.256	0.253
Area of primordium at initiation relative to apical dome area ($2 \log_e r$)	0.987	1.106

diffusion–reaction field by the application of substances which interfere with it. Of the plant growth substances, auxins and auxin analogues are the substances which most consistently affect leaf initiation and phyllotaxis when applied to apices. Indole acetic acid (IAA) caused irregular phyllotaxis and the formation of multiple leaves in *Tropaeolum* apices (Ball 1944), as did application of 2,4-dichlorophenoxyacetic acid (2,4-D) to *Phaseolus* apices (Furuya 1958, Soma 1968). Auxin applied to *Lupinus* and *Epilobium* apices caused enlargement of primordia and fusion of successive primordia, and also tended to result in displacement of subsequent primordia towards the site of auxin application (Snow and Snow 1937).

When triiodobenzoic acid (TIBA, an auxin transport inhibitor) was fed through the freshly cut basal end of the stem of young *Chrysanthemum* cuttings bearing about eight leaves their phyllotaxis was changed from spiral to distichous (Schwabe 1971). The main effect of the TIBA was to alter the shape of the apical dome, decreasing the apical angle (Table 7.2) and so making it narrower, and increasing the vertical spacing between primordia as well as decreasing the distance directly through the apical tissues from one primordium to its next neighbour. Primordium size was not appreciably affected. The interpretation was that by narrowing the apex, primordium initiation was being influenced to a greater extent than before by the previous primordium, so that the next primordium arose as far away as possible (opposite) the last-formed and so making the arrangement distichous. This is all consistent with a diffusion–reaction mechanism being involved and being affected by an auxin antagonist (TIBA). The results of this experiment presumably would also be consistent with the narrowing of the apex altering the stress pattern on the surface so that the phyllotaxis changed, but it is more difficult on the biophysical model to predict that the phyllotaxis would become distichous.

How TIBA acts at the cellular level to alter apex shape is not clear, apart from presumably inhibiting basipetal auxin transport. TIBA caused a lengthening of the plastochron in *Chrysanthemum*, consistent with the primordia being slightly larger at initiation (Table 7.2) but initiated slightly less frequently. This would be consistent with no change in the mean growth rate of the apex, which was what was deduced from the plastochron ratios (Table 7.2). It is not known whether the incidence of periclinal divisions, or the plane of division in general, in the apex was altered by this treatment.

Other auxin antagonists have also been shown to affect phyllotaxis (Meicenheimer 1981). *N*-1-naphthylphthalamic acid (NPA) is a potent inhibitor of auxin transport. When it was applied in a lanolin paste to one of the pair of youngest primordia this primordium became extended laterally, occupying more than half of the apical circumference (compared with one-half in the bijugate controls), and the apex then went on to produce primordia in a spiral instead of the opposite, bijugate, arrangement in the untreated plant. In the NPA-treated plants the plastochron was half that in the controls (which produced a pair of leaves per plastochron) but the relative growth rate of the apex was the same. What appears to have happened is that the NPA increased the proportion of the apical dome used to produce a primordium by preventing the diffusion away of auxin produced by the young primordium and instead allowing it to be available for promotion of extension of the primordium.

Similar results were obtained by the use of α-4-chlorophenoxyisobutyric acid (CPIB), an antagonist of auxin action (Meicenheimer 1981). This substance when applied to one of the youngest primordia completely or partially inhibited its growth, thus allowing the next primordium to occupy more than half of the apical circumference and so leading to spiral phyllotaxis. It can be seen that by either promoting the growth of a primordium or inhibiting it, the result was the same, that either that or the next primordium originated from more than half of the apical circumference so that subsequent primordia were formed singly, and in a spiral, instead of in pairs and opposite each other. In both cases the new divergence angle was indistinguishable from the Fibonacci angle of 137.5°, and the shape of the apical surface was changed, becoming less elliptical and so with less eccentricity (0.57) than in in the controls (0.75) (where 0 represents circularity, and increasing eccentricity gives values increasing towards unity; Meicenheimer 1981).

2,4-D has also been shown to alter primordium form in some plants. Applied to *Phaseolus* shoot apices (Pereira and Dale 1982) it caused the formation of a collar round the apex, apparently the result of a continuous primordium being formed right round the apex. This is consistent with 2,4-D blocking auxin transport away from the young primordium, increasing auxin concentration on the flanks of the apex and so causing the primordium to spread round the apex: a more extreme form of Meicenheimer's NPA experiment. Since auxin is known to increase the plasticity of epidermal cell walls (Kutschera, Bergfeld and Schopfer 1987) it is plausible that a locally increased concentration of auxin normally causes local increased wall plasticity and so the localized formation of a primordium, but that if auxin concentrations are increased all round the apex when 2,4-D is present, then wall plasticity is increased all round the apex and a collar forms.

These results are consistent with the idea that the newly formed primordia are sources of auxin required for their growth and establishment of their lateral extent, and that young primordia act as inhibitors of the formation of new primordia until they are sufficiently distant on the apical surface. This is all consistent with a diffusion–reaction mechanism for the initiation of primordia. It would be extremely interesting to know how the surface microstructure of the apex changes in these experiments.

Gibberellic acid may also affect phyllotaxis, but apparently by increasing the size of the apical dome without increasing primordium size at initiation, so that partitioning is altered and the plastochron ratio decreases (Maksymowych and Erickson 1977).

7.7.2. *Surgical and mechanical modification of phyllotaxis*

Experiments to try and provide evidence for the first available space theory, rather than a field theory for primordium positioning, were done by Snow and Snow (1948). Cuts close together, in the apical surface of *Lupinus*, restricted the apical surface available for a primordium (Snow and Snow 1952). Cuts encroaching on a potential primordium site could alter the position of a primordium so long as they were made more than half a plastochron before the primordium appeared (Snow and Snow 1933). The results were in general consistent with there being a minimum

space required for a primordium to form and with the primordium forming in the next space available to it. However, in a few cases, Snow and
Snow had to postulate that new primordia overlapped each other. Clearly
this is not consistent with each primordium occupying a distinct space
and is more consistent with interacting fields. Also, the cellular basis for
definition of a space is highly unlikely to depend on the mechanical
packing properties of distinct groups of cells. Presumably the available
spaces must be determined at the cellular level by chemical interactions
between cells. Their claims, that their experiments could discriminate
between packing and diffusion–reaction theories, appeared to rest on
their assumption that all the existing primordia would affect the position
of a new primordium if the primordia were repelled by some chemical
influence, but that if a primordium arose in the first available space only
the adjacent primordia would determine its exact position (Snow and
Snow 1948). However, this basic premise would no longer hold if the
repulsive effect of primordia changes, decreasing as they age after initiation. Also, incisions in the apex would surely disrupt the action of a
diffusion–reaction system and so lead to the same sort of results as
were found.

Similarly, experiments in which phyllotaxis was altered by bisecting
the apex do not allow discrimination between a diffusion–reaction or a
physical mechanism for phyllotaxis. In *Epilobium* and in four species of
Labiatae, in which the primordia are opposite and decussate, bisecting
the apex caused the phyllotaxis on each half to become spiral in some of
the plants (Snow 1942). Also, in *Euphorbia* incisions on the apex induced
spiral, instead of the usual decussate, phyllotaxis (Soma 1958). These
experiments imply that primordium positioning is not controlled in a
simple way by genes, but that it is a secondary outcome of the initiation
of primordia and can be modified by the size, configuration and activity
of the apical meristem.

Phyllotaxis in the sunflower inflorescence head could be disrupted by
isolating the centre of the apical dome from the rest of the apex by a
circular cut (Hernández and Palmer 1988). New primordia that formed
on the central isolated part of the apex were not ordered but apparently
randomly distributed, implying that the positioning of the newer primordia depended on the established pattern and that the centre of the apex
was not pre-patterned. However, this does not allow us to decide whether
the patterning is dependent primarily on mechanical or chemical factors.

Hernández and Green (1993) constrained a developing sunflower head
in a clamp so that growth was restricted to one axis instead of being

radial. Not only was the phyllotaxis altered but also some of the floret primordia developed only as bracts and did not form floral organs. Although this showed that mechanical constraint could affect development, it is not clear whether this was a direct biophysical effect of surface stress on the orientation of microfibrils, or the result of mechanically induced chemical changes, such as a stimulation of ethylene production or changes in cytosolic Ca^{2+} concentrations (Haley et al. 1995).

7.8. Chemical versus biophysical controls in positioning of primordia

Buds and leaves can be initiated on the *Graptopetalum* residual meristem by placing a drop of cytokinin on the meristem surface (Grayburn, Green and Steucek 1982). Presumably the biophysical changes associated with leaf formation then took place. It would be of great interest to know the sequence of biophysical events in the outer cell walls in this and other experimental situations where buds can be induced *de novo* (e.g. Tran Van Thanh 1981, Selker and Lyndon 1996) and when growth regulators can alter the size of primordia and so alter phyllotaxis (e.g. Meicenheimer 1981).

The only analysis of plastochron ratios and divergence angles in the developing flower, in *Silene*, showed that the positions of the sepals (after the first two) and the petals, sepals and carpels was best predicted on the basis that successive whorls were positioned with reference to the preceding whorl (Lyndon 1978b). The diffusion–reaction mechanism could account only for a less accurate positioning, although sufficiently so to prescribe the region of the apex where the next primordium would appear. It was therefore suggested that there were two sets of factors involved in positioning. One would be the same sort of inhibitors produced by leaf primordia which prevent the initiation of new primordia close by, and the other would be factors emanating from the existing floral organs which would promote the formation of the next whorl. In the light of Green's work, we can now propose that this second set of factors is not chemical but biophysical.

Green (1994, p.1786) concluded that 'chemical activation determines whether primordia will form, physics determines where'. He implied that if there is any localization at all of chemical activation, even imprecisely positioned, it will also be involved in determining primordium positioning. In the *Silene* flower, the positioning of the primordia of one whorl clearly depended on the positions of the primordia of the preceding whorl, which could be the result of the biophysical configuration of the apical surface. However, the sequence of initiation of primordia was best predicted by assuming that primordia inhibited the initiation of new primordia, but only for two plastochrons (Lyndon 1978b); this would be consistent with the production of an inhibitor by newly initiated primordia as in the field theories. It seems that in the flower, chemical factors could determine the sequence (whether primordia will form) and could determine roughly where, but precisely where could depend on the biophysical structure of the apical surface.

The problem about the diffusion–reaction hypothesis for the formation and positioning of primordia is that there is no direct evidence of the chemical morphogens that would be involved. Methods for measuring in the shoot apex the changing concentrations of small molecules, such as plant growth substances, are needed. At least with the biophysical theories there is evidence of the changes in microfibril orientations in the epidermal cells. But the hypothesis, that primordia normally form by a buckling of the apex surface, is not consistently supported by direct evidence. Holes pricked in the surface of an apex often tend to expand rather than close up, suggesting the surface is often under tension rather than compression (Selker, Steucek and Green 1992). Although this may be partly because the apex dries out very quickly in air, when cuts were made in apices held under water or solutions of various concentrations of mannitol, as an osmoticum, the cuts invariably gaped in *Euphorbia* and tomato apices. Only in pea did the cuts close up. This showed that the apical surfaces in *Euphorbia* and tomato were under tension but under compression in the pea (Hussey 1973). This is consistent with the rates of growth and cell division in the epidermis of the incipient primordium being less than in the underlying cells in tomato apices (Hussey 1971b) but the same in the pea (Lyndon 1971). It does not seem consistent with buckling of the surface being a general mechanism for primordium formation.

A possible problem for any biophysical hypothesis that implicates microfibril orientations as cause rather than result of organogenesis is that mutants of *Arabidopsis* have been found in which microtubule orientation is random in most cells and yet the plants still form leaves, albeit abnormal ones (Lloyd 1995, Traas et al. 1995). It is not known what the microfibril orientations are in these cells. Are surface microfibrils still aligned by stresses imposed by developing organs? Are microfibrils more concerned with determining organ shape rather than organ initiation? This indeed seems probable (Green, Steele and Rennich 1996). Obviously more information is needed before the relative contributions of diffusion–reaction mechanisms and biophysical mechanisms to primordium initiation can be evaluated. It would be surprising if both were not involved, one complementing the other. It seems likely that a diffusion–reaction mechanism, perhaps involving auxin, is involved in primordium initiation and is sufficient roughly to determine the position of a primordium, but that precise positioning and particularly the shape of the primordium is dependent on biophysical factors.

7.9. Position and sequence

Usually the primordia form in acropetal sequence on the apex so that the position of a new primordium is a function of when it is initiated. However, there are some apices in which position and sequence are apparently determined quite separately. In many leguminous flowers the primordia do not form in developmental sequence but initiation starts at the back of the apex and progresses in a wave to the front (or vice versa in some species) so that primordia in different whorls and at different positions in the acropetal sequence are initiated simultaneously (Tucker 1984a). The biophysical structure of the surfaces of these apices and whether they are prepatterned, or whether the biophysical structure forms as the primordia form is not known. These flowers seem to show that there are separate mechanisms for determining the pattern (biophysical?) and the sequence (diffusion–reaction?), operating sequentially, whereas in most other flowers they operate simultaneously. However, the operation of these mechanisms seems so different in these flowers that it is difficult to see how primordium initiation and positioning in these leguminous flowers can be adequately explained by any hypotheses so far proposed.

7.10. Genes for phyllotaxis and primordium initiation

The same plant can have changing phyllotaxis during its vegetative development and the change from opposite to spiral as the plant matures is quite common. Surgical experiments and the application of chemicals, especially growth substances, can change phyllotaxis. These observations suggest that the precise phyllotactic pattern is not prescribed by the genes, but that the system for generating it is. Genes that alter phyllotaxis may therefore be expected to be those that alter apex or primordium size, or growth substance concentration within the apex, or change cell wall characteristics which affect microfibril properties and orientation. Analysis of the function of such genes should throw light on the mechanisms of primordium formation and positioning.

Genes affecting the relative sizes of primordia and apical dome are *ABPHYL*, which changes the proportion of apical cells that becomes committed to a primordium (Greyson and Walden 1972), and *CLAVATA* and *FASCIATA*, which increase the size of the apical dome but apparently not the incipient primordia, since phyllotaxis is disturbed although the divergence angle remains unchanged (138°) (Leyser and Furner 1992, Clark, Running and Meyerowitz 1993). The *fuf* (fully fasciated) mutant in *Arabidopsis* doubles the diameter of the meristem and alters the phyllotaxis (Medford et al. 1992).

If phyllotaxis is an emergent phenomenon, which is the unspecified outcome of the processes of primordium formation, then genes for phyllotaxis as such will not exist. The genes controlling the underlying processes that result in, but do not specify, pattern must be searched for. It needs to be asked whether, in fact, the cyclic formation of new organs during normal iterative growth involves changes in gene expression. If the supposed activators and inhibitors of a diffusion–reaction mechanism regulate the activity of enzymes directly, then there is no need to involve the genes directly in this regulation: the genes will have set up a self-regulating system. Only if regulation is assumed to involve transcriptional or translational events does the problem pass from the realm of cellular biology to molecular biology. The formation of primordia appears to be a cyclic activity at the metabolic level, analogous to the cell cycle (Ormrod and Francis 1993) and perhaps transcriptionally controlled only at very few points, or only one. Changes in gene expression could be mainly involved in developmental switching, e.g. at the initiation and differentiation of the leaf itself, but not in most of the cyclic changes in the shoot apex which lead up to primordium initiation. Mutants for primordium formation need to be looked for in the embryo. One example is the 'reduced' form of embryos homozygous for the *lanceolate* mutant of tomato, which does not form either primordia or a shoot apex (Caruso 1968).

Partitioning the Apex: the Size of the Apical Meristem and the Primordia

8.1. Plastochron

The repeated production of primordia reflects the cyclic functioning of the apex and the iterative formation of the modules that make up the aerial part of the plant. The formation of a primordium, with its associated axial tissue, represents the partitioning of the apical tissues between the newly formed tissues and the apical dome which remains, and which then repeats the process.

The interval between the formation of successive primordia is the plastochron, a measure of developmental time. In plants that produce a pair of primordia simultaneously or a whorl, the plastochron is the interval between the production of successive pairs or whorls. Since it is the nature of a cyclic process that any point to delimit the beginning or end of the process must be arbitrary, a more general but no less precise definition of the plastochron is the interval between successive similar events at the shoot apex (Esau 1965).

8.1.1.   The events of a plastochron

The formation of a primordium is usually accompanied by other events which are not so immediately evident at the vegetative apex, namely the formation of the initials of the axillary bud and of the node and internode (see Chapter 2). In the flower, the formation of a primordium is accompanied by the formation only of the node initials – the axillary bud and the internode do not seem to be formed (Lyndon 1987a) (see Chapter 10).

8.1.2.   Length of the plastochron and the phyllochron

The length of the plastochron is usually measured as the reciprocal of the rate of primordium initiation and is often of the order of several days. A much more frequently used measure of development is the phyllochron, the interval between the visible appearance or emergence of successive leaves in the intact plant, the inverse of which is the rate of leaf appearance. The plastochron and phyllochron are not necessarily the same interval. The rate of primordium formation and the rate of leaf appearance are the same only if the period between leaf primordium formation and visible leaf appearance is constant. In many cases it is not. In wheat, for instance, leaves emerge at the rate of only 0.42 per plastochron, so that a phyllochron equals 2.38 plastochrons (Fig. 8.1) (Hay and Kemp 1990). This implies that an increasing number of young primordia are 'piling up' at the apex because they are developing slower than they are being produced. This also happens in other grasses (Evans 1960) and, less dramatically, in the pea (Lyndon 1977a). In *Cyclamen*, the phyllochron is longer than the plastochron during the first half of the growing season but is shorter during the second half as the rate of leaf initiation decreases and the plastochron therefore lengthens (Sundberg 1982). The rate of primordium formation greatly outstrips the rate of leaf and scale maturation during bud set, especially in conifers (e.g. Cannell and Cahalan 1979), and the opposite happens during bud break when leaves enlarge and unfold very rapidly in the spring.

The plastochron index (Erickson and Michelini 1957) is extremely useful for defining developmental stages, but it must be borne in mind that this is essentially a phyllochron index. The definition of the plastochron given by Maksymowych (1990, p.6) as 'the interval between corresponding states of development of successive leaves at their initiation, maturity, or any stage of development which can be used as a reference

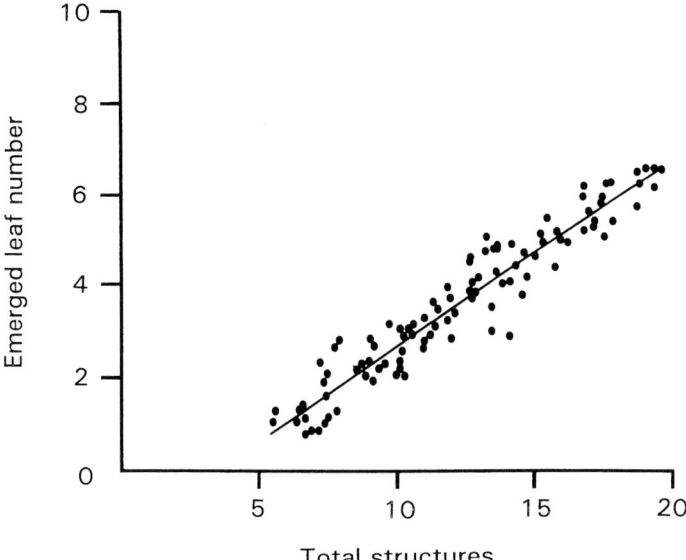

Figure 8.1 Relationship between emerged leaf number and the number of total structures (primordia on the apex plus emerged leaves) at the mainstem apex of wheat (*Triticum*). Leaves are produced faster than they develop so that primordia 'pile up' on the apex. From Hay and Kemp (1990).

stage' is only valid if the rate of development of every leaf is shown to be identical. What Maksymowych defines is, in fact, the phyllochron which, as we have seen, is not necessarily the same as the plastochron. Perhaps the plastochron index should be renamed the phyllochron index.

8.1.3. Plastochron: developmental cause or consequence?

In models of apical functioning (e.g. Thornley 1975a, Charles-Edwards et al. 1979, Veen and Lindenmayer 1977, Thornley and Cockshull 1980) the plastochron is usually thought of as the resultant of the processes that control leaf initiation, i.e. it is determined by the component processes rather than being a rhythm in its own right, like a circadian rhythm. If it were, then one might, for example, expect to see variation in the size of primordia at initiation or the positions at which the primordia are formed. Since primordium size at initiation tends to remain constant or to change only gradually or at particular identifiable times during development, e.g. on flowering, this points to size of primordia and the positions at which they are formed being under close control, and that it is these that determine, incidentally, the plastochron; or that the plasto-chron is determined by the same processes that determine size and pos-itions of primordia. If it were an intrinsic rhythm then we might expect it to be perturbed by inhibitors of rhythmic phenomena. Applications of verapimil (a calcium-blocking drug that can affect physiological rhythms) to pea shoot apices had no detectable effect on the plastochron (R.F. Lyndon, unpublished data); but, of course, a single, negative experiment tells us little.

The general consensus seems to be that the plastochron is the conse-quence of the interaction of the size of a primordium at initiation and the rate at which the apex expands radially to create sufficient space at a minimum distance from the apical summit in order for the next primor-dium to form. The plastochron could perhaps be a rhythm resulting from the production and degradation of an inhibitor, of primordium initiation in the apex, which also is concerned with determining primordium size. Although the plastochron may only be the resultant of several processes, it may none the less apparently act as a determinant of further develop-ment. This may be because it is correlated with, or depends on, the same factors that do determine further development. An example of this is in wheat. Rate of leaf emergence in wheat is a function of the number of total primordia formed up to collar formation and therefore is not affected directly by the environment. The determinant of further devel-opment appears to be the plastochron, or at least the processes which in turn determine it. Environmental effects must be working through their effect on primordium initiation (Hay and Kemp 1990).

8.2. Partitioning the apex: redefinition of the apical dome each plastochron

The size and shape of the shoot apex are often characteristic of a parti-cular species or of related species. For instance, the apices of palms are large, shallow domes, usually about 1mm in diameter. Apices of *Helianthus* species are usually disc-like, about 100μm across, with pri-mordia on their rims. On the other hand the pea apex is hemispherical, with the primordia being formed on its sides. The general size and shape of an apex must be determined genetically, although this may not be via only one or two genes, but as the result of the action of many genes concerned with the processes contributing to the functioning of the apex.

The shoot apex may change in size and shape as the plant develops during ontogeny, and especially when it flowers. There is a regular cyclic change in the size of the apex as it grows in the course of successive plastochrons. The apex grows from minimal area, when a primordium has just been initiated, to maximal area, just as the next primordium is about to be initiated. This plastochronic variation in apical surface area is least in those apices which form relatively small primordia singly, and greatest in apices forming relatively large single primordia (as in the pea) or pairs or whorls of primordia. Since some of the tissues of the apical dome are incorporated into the new primordium each plastochron, then the apical dome is redefined each plastochron. The apical dome after primordium initiation is a new structure which consists of only a fraction of the dome from the previous plastochron. But the cells forming the apical dome of minimal area of one plastochron are the same cells (plus their immediate descendents) that form the maximal area one plasto- chron later. When a primordium is then formed, the size of the apical dome of new minimal area obviously depends on what proportion of the apical dome is used up in the formation of the primordium.

The growth of the apical dome through successive plastochrons can be shown as a saw-toothed line (Lyndon 1977b) (Fig. 8.2). The growth of the apical dome is assumed to be exponential (see Chapter 3), and this is represented in Fig. 8.2 by the upwardly sloping linear sections of the graph. Each plastochron the formation of a primordium and its asso- ciated tissues results in an abrupt reduction in volume of the apical dome, from maximal to minimal. The size of the reduction corresponds to the volume of the tissues forming the primordium and its associated axial tissue. Because the scale on the y-axis is logarithmic, the length of each vertical line is directly proportional to the size of the primordium (and its associated axial tissue) relative to the apical dome. If the vertical line (Fig. 8.3) becomes shorter each plastochron this implies a relatively smal- ler primordium, and if longer, a relatively larger primordium. The mean size of the apical dome each plastochron is represented by the midpoint of each of the sloping lines. A line joining these midpoints will remain level if the apical dome size remains constant from plastochron to plas- tochron. However, a rise in this line implies a gradually enlarging dome, and a fall, a gradually diminishing dome. If the growth of the apex during a plastochron is exactly compensated by the formation of a new primor- dium then the apical dome will remain constant in size. However, if each new primordium is formed before or after the precise amount of necess- ary growth has occurred, then the apex will shrink or enlarge respectively.

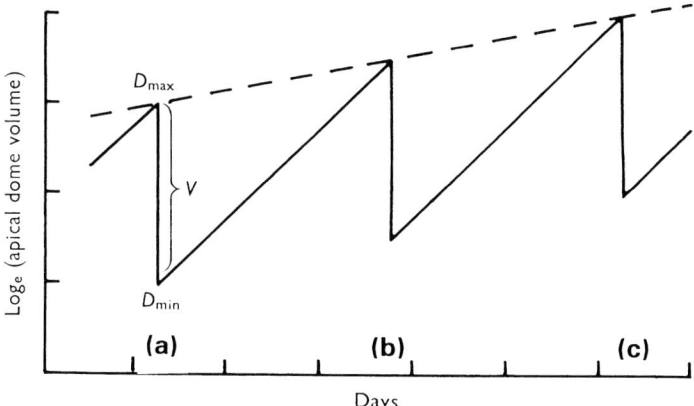

Figure 8.2 Growth of the apical dome in successive plastochrons. The apical dome grows exponentially (the upward sloping solid lines) and every time a primordium is initiated (a,b,c), the dome size is reduced as some of the tissues are partitioned off into the primordium, and the dome is reduced in size from maximum (D_{max}) to minimum (D_{min}). The plastochron is the interval between a and b, or b and c. V represents the size of the primordium, and the amount of tissue that must be made in the next plastochron for dome size to be maintained. If, as in this diagram, the plastochron is longer than required, and the size of the primordium relative to the apical dome remains constant (i.e. V remains constant) then the apical dome will gradually increase in size (dashed line) even though the growth rate (the slope of the line between a and b, and b and c), remains constant. From Lyndon (1977b).

This will occur even if the growth rate of the apex and the proportion of tissue cut off to form the primordium remain constant from plastochron to plastochron. In many plants, the apex gradually enlarges during vegetative growth, because the amount of growth during a plastochron is more than is required to form a primordium of constant relative size. In the flower the apical dome eventually disappears because the proportion used to form the floral primordia finally increases enormously, as the growth of the dome becomes insufficient to maintain the apical dome as well as continuing formation of primordia of a constant size.

The apical dome may therefore remain constant in size or change in size steadily, depending on the relative size of primordia on initiation. If there is a change in the growth pattern of the apex this can only be for one reason: that the proportion of tissue that is cut off each plastochron (the height of each vertical line in Fig. 8.2) has changed because of a change in the size of the primordium relative to the apical dome. This could occur in three ways. The primordium may change in absolute size, without a change in size of the apical dome and without a change in growth rate. Second, the growth rate of the whole apex (the slope of the inclined parts of the graph) could change, but without a change in the plastochron or the absolute size of the primordia, so that the size of primordia relative to the dome changes. Third, the apical dome could increase in size by a transient increase in growth rate between the formation of successive primordia. Theoretically a fourth possibility is that the plastochron could change in length (the horizontal distance between vertical lines), without necessarily a change in apical growth rate or in the absolute size of a primordium at initiation. However, this would be apparent as a change in the relative size of a primordium on initiation. It is therefore difficult to distinguish between a mechanism that determines the size of the primordium at initiation relative to the apical dome and one which determines the length of the plastochron directly. If the plastochron were determined independently then a much greater plastochron-to-plastochron variation in the size of the apical dome and the primordia would be expected and there is little or no clear evidence for this. Fig. 8.3 shows how the same outcomes, of constant or changing apical dome sizes, can result from different combinations of growth rate and partitioning. If the relationship between these two parameters changes, then the apex will change in size. The fact that this relationship can change, and is not always constant, implies that primordium size at initiation is not controlled entirely by apical growth rate. This will become more apparent when examples of changes in partitioning during development are examined.

8.2.1.   Measurement of partitioning: the relative sizes of primordia and apical dome

In order for partitioning to be quantified it is necessary to measure the relative sizes of primordia and apical dome. The amount of tissue assigned to the primordium and the amount assigned to the apical dome need to be known. This can be measured as volume or area. Volume can be measured from serial sections as the amount of tissue that is made by the apex during a plastochron and is incorporated into the new primordium and its associated stem tissue but not into the remaining apical dome. Ideally this value should be known at the moment the primordium is formed. To do this it is necessary to define the primordium and the apical dome at the moment of initiation. Such

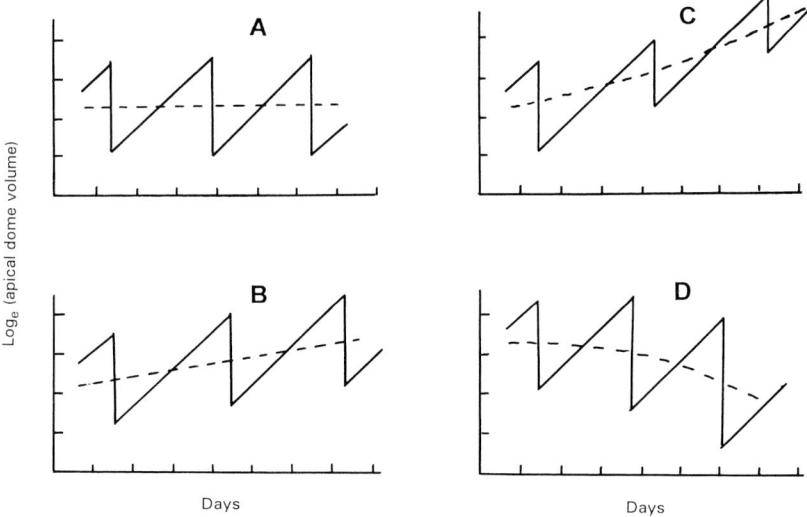

Figure 8.3 Different growth patterns for the apical dome. For explanation of lines see legend to Fig. 8.2. The growth rate (the slope of the upwardly sloping lines) is the same in A, B, C and D. (A) The growth rate is exactly that required to reconstitute the apex in the course of a plastochron; primordial size, absolute and relative, is constant, and so mean apical dome size (dashed line) remains constant. (B) The plastochron is longer so the apical dome increases in size. (C) The tissue is partitioned less towards the primordium each plastochron so that the primordia are becoming relatively smaller and the dome larger. (D) Partitioning is biased towards the primordium, which is becoming relatively larger each plastochron, so that the apex is gradually getting smaller.

measurements, from serial sections, of the sizes of primordia at first appearance have been done for *Silene* (Lyndon 1977b).

The other way to measure primordium and apical dome size is from their relative areas. These can be deduced from the plastochron ratio (Richards 1951). This is the ratio of the distances (R_1 and R_2) of successive primordia from the apical centre, and so

$$\text{plastochron ratio, } r = R_1/R_2.$$

Assuming constant partitioning, the area increase for each plastochron between minimum apical area (when a primordium has just been initiated) and maximal apical area (when the next primordium is just about to be initiated) is equivalent to the area of the shaded annulus (Fig. 8.4). This is equal to:

$$\pi r^2 - \pi = \pi(r^2 - 1)$$

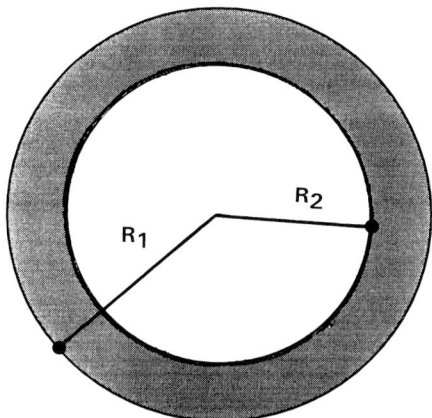

Figure 8.4 Diagram of an apical dome viewed from above. A primordium is initiated at a radius R_2 from the centre of the apex. After one plastochron's growth this primordium will now be at a radius of R_1. The shaded annulus between R_1 and R_2 represents the area of tissue that must be generated for a primordium to form.

and relative to the minimal area of the apical dome (π) is:

$$\pi(r^2 - 1)/\pi = r^2 - 1,$$

which therefore represents the increase in area of the apical dome each plastochron, relative to the minimal area of the apical dome.

Since the primordia are measured from their centres, and radial growth is exponential, Richards (1951, p.553) explained that 'the mean area [of the apical dome] is given by the ratio of the difference between the extreme areas to the difference between their natural logarithms, i.e. mean apical area is proportional to ... $(r^2 - 1)/(2 \log_e r)$. Since primordium area is similarly proportional to $r^2 - 1$, the ratio of mean apical area to primordium area $= 1/(2 \log_e r)$'. The reciprocal is the ratio of primordium area to mean apical area, $2 \log_e r$, and so this is a measure of the area assigned to a primordium half a plastochron after its first appearance, and can be taken as a measure of the area of a primordium at initiation, relative to the apical area. The apical area is the cross sectional area of the apex at the level of insertion of new primordia. Thus, measurement of the plastochron ratio, r, allows an estimate of the area of a primordium at initiation relative to apical area to be obtained. If the apical angle is also known (Fig. 8.5) then the actual area of the primordium on the apical surface relative to the apical area can be calculated, and if the cross-sectional area of the apex is known then these can be translated into absolute values (e.g. Lyndon 1978b).

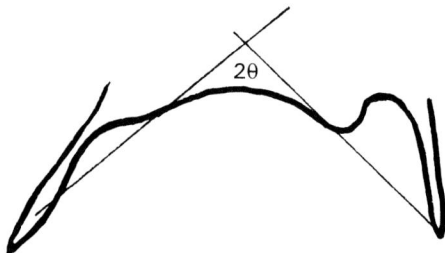

Figure 8.5 The apical angle (2θ) is that subtended by lines tangential to the apical surface where the most recent primordia have been formed, i.e. at the base of the apical dome.

8.3.   Relationship between plastochron and cell doubling time

A small apex producing large primordia is more likely to contribute cells to the apical flanks each plastochron than a large apex producing small primordia. The restitution of the apical dome in the latter will require relatively fewer cells as a smaller proportion will have been displaced into the zone of primordium initiation (Dulieu 1969). The doubling time of the apex in plastochrons will therefore be linked to the size of the primordia at initiation relative to the size of the apical dome.

Although measurements derived from the plastochron ratio are confined to the apical and the primordium surface, Richards (1951) pointed out that if it is assumed that the vertical growth rate in the apex is the same as the radial rate, then the relative growth rate in volume per plastochron is $3 \log_e r$. Since the doubling time is the number of plastochrons taken to double the volume when the relative rate of volume increase is $\log_e 2$, then if the plastochron ratio, r, and the length of the plastochron are known, the apical doubling time can be estimated:

Apical doubling time (plastochrons) $=$ number of plastochrons$/\log_e 2$

and so

Apical doubling time (h) $= [(\log_e 2)/(3 \log_e r)] \times$ plastochron (h).

If cell volume remains more or less constant in the apex, then the apical doubling time is also the mean cell doubling time and gives an estimate of the length of the cell cycle in the apex, if it is assumed that all the cells are cycling. This relationship has similarly been shown by Jean (1994) who, by expressing the area ratio, $1/(2 \log_e r)$, as a function of the number of parastichies was able to derive an equation relating the plastochron directly to the phyllotactic pattern and the growth rate of the apex.

It is clear that the relationship between the plastochron and the cell doubling time depends on the value for 3 log $_e r$ equivalent to the size of the primordium at initiation. The larger the primordium at initiation, the longer the plastochron has to be relative to the doubling time in order to prevent the apical dome shrinking in successive plastochrons. On the other hand, if the primordium is very small, then the plastochron could be quite short even though the cell doubling time is very long. The plastochron can be longer or shorter than the doubling time depending on the size of the primordium on initiation. In apices with a low order of phyllotaxis, such as distichous or decussate, where the plastochron ratio, r, is relatively large, the doubling time is shorter than the plastochron. In the pea, the doubling time was 0.63 plastochrons (Lyndon 1968a), and in wheat, 0.5 plastochrons (Evans and Berg 1971). In apices with a higher order of Fibonacci phyllotaxis, with small r and small primordia, the doubling time may be much longer than the plastochron, as in the oil palm, where the doubling time was 2.82 plastochrons (Rees 1964).

8.4.   Constant and changing partitioning during growth and development

Partitioning has been followed in several species during vegetative growth. In *Lupinus* (Sunderland and Brown 1976) and *Silene* (Lyndon 1977b), the apex gradually enlarges with age and so do the sizes of the primordia at initiation. The growth that is surplus to requirement for formation of a new primordium of the same size as in the last plastochron, is added proportionately to the primordia and the dome and the partitioning is maintained. This is also what seems to happen in maize during vegetative growth as the apex enlarges (Abbe, Randolph and Einset 1941, Abbe and Phinney 1951, Abbe, Phinney and Baer 1951). The partitioning in maize shows the same pattern as in Fig. 8.2, but this is not perturbed by a rapidly increasing overall growth rate, and a correspondingly shorter plastochron, as the apex develops (Abbe, Phinney and Baer 1951). The *Agropyron* apex increases five-fold in size from plastochrons 4 to 11 and then decreases to half this size in the next four plastochrons (Smith and Rogan 1979). Although this is accompanied by a reduction of the relative growth rate to a tenth of its original value, there is a corresponding lengthening of the plastochron from 3 to 20 days, because the partitioning remains essentially unaltered. The cell doubling time in the apical dome is about 0.47 plastochrons and the primordium is always about 44% of the cells available when the apex reaches maximal size at the end of each plastochron (Smith and Rogan 1979).

8.4.1. Changing partitioning: how is it brought about?

In *Ambrosia* the phyllotaxis changed from decussate to spiral, at first with 2:3 parastichies and then to 3:5 as the apex enlarged (Soma and Kuriyama 1970). The apical dome increased in mean diameter from 53 to 130μm, while the size of the primordia at initiation, measured as their tangential extent in cross section, remained apparently constant at about 96μm. The plastochron ratio also showed no measurable change, although some change would have been expected. Soma and Kuriyama note that in *Chenopodium* the plastochron ratio was found by Gifford and Tepper (1962) to decrease from 1.087 to 1.049 as the phyllotaxis changed from 2:3 to 3:5 parastichies. In *Ambrosia* the change in phyllotaxis from opposite, decussate to spiral was therefore associated with an enlarging apical dome but a constant size of primordium at initiation.

In other plants, too, the apex can be observed to enlarge as the plant grows, e.g. in *Helianthus tuberosus*, as the phyllotaxis changes from decussate to spiral. This is consistent with the absolute size of a primordium at initiation remaining constant while the apical dome enlarges, surplus growth being added disproportionately to the apical dome. This is also what happens in *Xanthium* treated with gibberellic acid (Table 8.1) (Maksymowych, Cordero and Erickson 1976). The growth rate is essentially the same before and after treatment, with a doubling time of about 60h, but the apical dome has enlarged to about twice the size of that in the untreated controls although the sizes of primordia at initiation remain unchanged so that the partitioning has changed, as shown by reductions in the plastochron ratio from 1.377 to 1.185, and in the relative area of a primordium on initiation (2 log $_e r$) from 0.64 to 0.34 (Maksymowych, Cordero and Erickson 1976).

A similar change in partitioning commonly occurs at the transition to flowering, when the apical dome enlarges (see Section 8.4.3). In *Silene* this was shown to be brought about by an increase in growth rate of the apical dome for only one plastochron, just prior to the transition

Table 8.1. *Effect of gibberellic acid on apical growth in* Xanthium.

	Untreated controls	GA$_3$-treated
Plastochron (days)	3.3	1.9
Plastochron ratio	1.377	1.185
2 log$_e r$	0.64	0.34
Apex doubling time (h)	57.1	61.9
Mean apical diameter (μm)	153	206
Apical cross section area (μm^2)	18148	33333
Apical volume $(2/3\ \pi r^3)$ (μm^3)	919,508	2,288,899
Absolute area of primordium at initiation (μm^2)	11614	11333

Calculated from the data of Maksymowych, Cordero and Erickson (1976).

(Lyndon 1977b). Presumably gibberellic acid (GA$_3$) in *Xanthium* has a similar transient effect on growth rate of the apical dome. A transient increase in growth rate of the apex can allow the apex to alter its partitioning, but may often pass undetected. As noted by Cannell (1976, p.553), although 'logistic and other functions can adequately describe long-term trends [they] are not good enough to determine short-term plastochrone changes as may occur, for instance, during the transition of pine buds from one cycle to another'.

Flowering is accompanied by an absolute reduction in the sizes of the primordia at initiation (see Section 8.4.3). In *Xanthium* treated with GA$_3$ this did not happen (Table 8.1), nor did the plants flower. The implication is that apical growth rate and primordium size are under quite different controls. The question then arises as to whether constant partitioning means that the two become linked, and if so, how? In constant partitioning, is the size of the primordium at initiation a function of the size of the apical dome, and if so, how?

8.4.2. *Growth rates of apical domes and what they mean*

In vegetative apices, the shoot apical dome often enlarges gradually with age, but in most plants, as in grasses (Evans 1960), enlargement of the apical dome is particularly characteristic of the transition to flowering (see Chapter 9). At first sight it appears as though the apical dome must be growing faster to enlarge but this is not necessarily so. How, then, are increases in apical size brought about?

The plastochron ratio can be used to derive $2 \log_e r$, which is a measure of the area of the primordium at initiation relative to the apical dome. Similarly $1/(2 \log_e r)$ is the area ratio (Richards 1951) and represents the area of the apical dome relative to the size of the primordium at initiation. Clearly the relative sizes of apical dome and primordium depend on how the apex is partitioned between them each plastochron. If the partitioning remains constant, then whatever the growth rate of the tissues, phyllotaxis will remain constant. If the growth rate increases and the absolute size of a primordium at initiation remains constant, then the apical dome also remains the same size at the beginning of each plastochron, but the plastochron shortens. If $2 \log_e r$ (i.e. partitioning) remains constant and the plastochron also remains constant as growth rate increases, this is because the absolute size of both primordium and apical dome are increasing. If the apical dome increases in size from plastochron to plastochron, but the primordia do not do so proportionately, then the increase in apical dome size is because the partitioning has changed and $2 \log_e r$ (and hence r, the plastochron ratio) has decreased. Measurements of rates of growth of the apical dome alone (measured as increase in size of the apical dome from plastochron to plastochron) are therefore almost meaningless (e.g. Jacobs and Pearson 1992). It is essential also to know the plastochron ratio.

Complete measurements that allow the growth to be properly interpreted have been made for conifer apices. In *Picea* seedlings growing in long days, the relative growth rate of the apex increased for about the first 3 months after sowing. This was accompanied by a reduction of the plastochron ratio (and so of the relative size of a primordium at initiation, $2 \log_e r$), and of the plastochron, so that the new tissue was partitioned more towards the apical dome than the primordia, resulting in an increase in the size of the apical dome and a doubling of its diameter (Cannell 1978). Later, in the summer, the relative growth rate declined and the plastochron lengthened, but with little change in the plastochron ratio (or of $2 \log_e r$) and so this resulted in the apical dome size stabilizing or reducing slightly (Cannell 1978). When seedlings were grown in short days to stimulate bud set (Cannell and Cahalan 1979), the increase in apical dome size was amplified (Fig. 8.6A) as the plastochron ratio (and $2 \log_e r$) decreased (Fig. 8.6B) at the same time as tissue relative growth rate increased (Fig. 8.6C), and so primordia were formed rapidly on an expanding apical dome. Eventually the growth rate declined, the plasto-

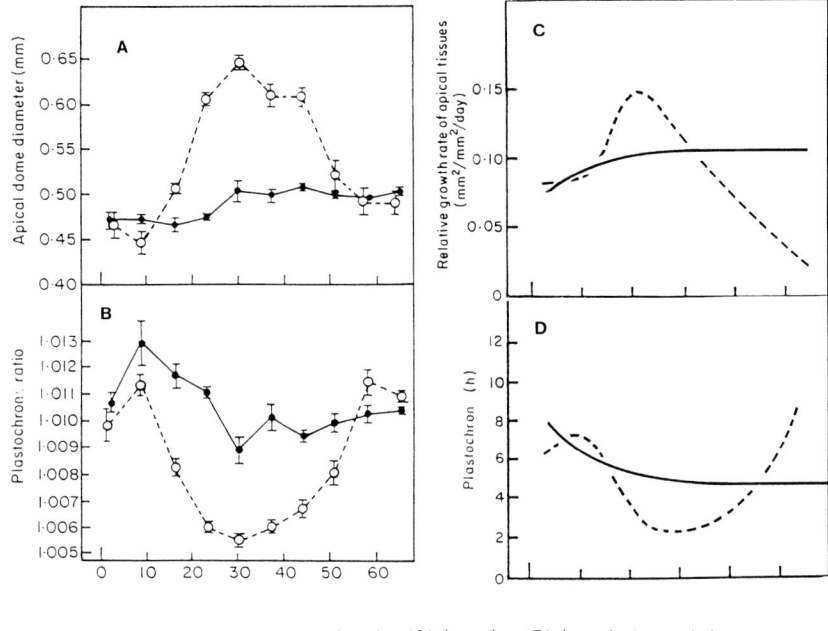

Figure 8.6 Growth of *Picea* shoot apices in short days (10-h photoperiods, - - -) and long days (17-h photoperiods, ——). (A) Apical dome diameter; (B) plastochron ratio; (C) apical dome relative growth rate; (D) plastochron. From Cannell and Cahalan (1979).

chron ratio increased, the plastochron lengthened (Fig. 8.6D), and the apical dome reduced in size to the long day control value as it became partly used up in the formation of primordia.

A survey of *Pinus contorta* and *Picea sitchensis* from differing provenances, including extremes of their natural ranges in northwestern Canada and USA, but all grown under the same conditions in Scotland, showed that the rates of primordia initiation were, in all cases, closely correlated with apical dome diameters (Cannell and Willet 1975). Since the maximum relative growth rates were essentially the same for apices of *Pinus* trees from all provenances (Cannell 1976) (Fig. 8.7), this implies that the size of primordia at initiation was probably fairly constant in absolute terms, so that larger apices make primordia faster simply because of the greater tissue mass available to generate new tissues. Calculation of primordium area in *Picea* from the data of Cannell and Cahalan (1979) shows that it indeed apparently remained about the same (0.003mm^2) despite an approximate doubling in apical area. If the cells each have a surface area of about 100μm^2, this would

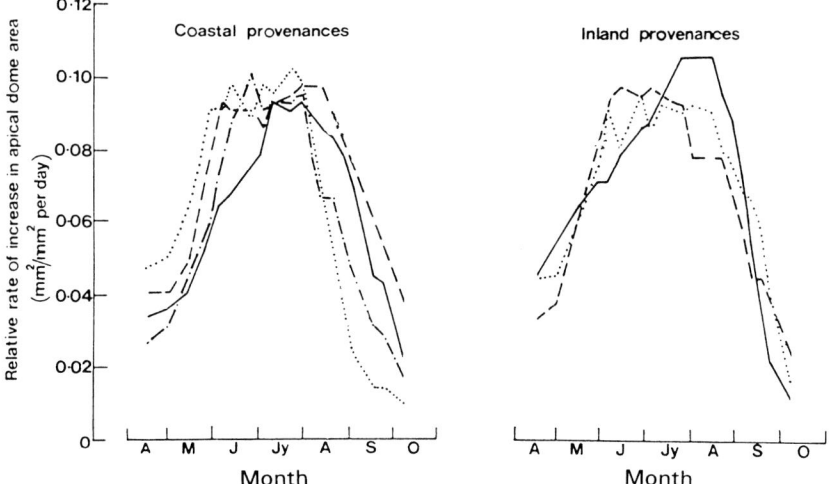

Figure 8.7 Relative rate of growth of the apical dome in *Pinus contorta* from seven provenances in northwest Canada and the USA, all grown under the same conditions in Scotland. The growth rate increases to a maximum in the summer and declines again in the autumn. From Cannell (1976).

imply about 30 surface cells involved in the initiation of each primordium, the same number as estimated in other plants from clonal analysis (see Chapter 6).

Similar measurements were done for apices of *Picea abies* (Gregory and Romberger 1972a, b, Romberger and Gregory 1977). They proposed the investment ratio (ϕ) as a measure of partitioning and defined it as 'that fraction of the total tissue volume production of the apical dome (ΣVG) that is retained by the dome, thereby increasing its capital volume by ΔV [the capital volume increment]' (Romberger and Gregory 1977, p.628); thus

$$\phi = \Delta V / \Sigma VG.$$

The investment ratio changed with age, being highest in small, young apical domes, and declining rapidly with age, reaching zero when maximum dome size was attained (Romberger and Gregory 1977). These findings are very similar to what was also found in *Picea sitchensis* (Cannell and Cahalan 1979), but expressed in a slightly different way. It is notable that in *Picea abies* all the changes in the apex, including the increase in apical dome volume and the shortening of the plastochron, took place while the relative growth rate of the apex as a whole was gradually slowing down so that there was a continuous increase in the volume doubling time. This illustrates very neatly how the way the apex is partitioned and the apical dome changes in size is not dependent on the rate of growth of the apex.

8.4.3. *Transition to flowering*

As plants make the transition from vegetative to floral there is often an increase in the rate of primordium formation (e.g. Kirby 1974; also see Chapter 9). This is almost certainly because of a reduction in the size of primordia at initiation (e.g. Snow and Snow 1952) relative to the apical dome, which enlarges. There is also reduced growth of the leaf primordia as shown by the reduction in size of the later-formed leaves and bracts just below the inflorescence (e.g. *Sinapis*, Bernier 1964; *Bryophyllum*, Zeevaart 1969; *Antirrhinum*, Bradley et al. 1996) and particularly by the reduced size of the bracts (compared with leaves) in the capitula of the Compositae.

On flowering, the apical dome enlarges (Evans 1960), so that the size of the primordia at initiation relative to the apical dome decreases. Eventually the opposite may happen, a gradual increase in the size of the primordia at initiation relative to the apical dome, either as a result of larger primordia, or a constant primordia but a shrinking dome. This is typically found in flowering meristems when the final flower primordia are formed on an inflorescence, such as in the capitulum of the sunflower. In this case growth will contribute disproportionately to the primordia, which tend to be a constant size on initiation, rather than to the apical meristem, which therefore shrinks.

The transition to flowering has been modelled for *Chrysanthemum*. Two main models have been proposed. In the first (Charles-Edwards et al. 1979), it is assumed that the inputs (assimilates, etc.) into the apical dome are constant. Since it has been observed that the *Chrysanthemum* apex makes the transition to flowering when the apex reaches a diameter of about 0.26mm, which is when the first bract is made, it was therefore assumed:

1. that when the apical dome reaches 0.26mm diameter the first bract is initiated;
2. that an inhibitor of primordium initiation is distributed through-out the apical dome, and that this inhibitor is produced by all primordia; and

3. that leaf primordia do not compete with the apical dome for assimilates, etc., but that bract primordia do.

This is consistent with the observation that leaves tend to be displaced away from the apex but bracts and floral primordia do not, because of the lack of internode development. This model simulates the changes in apical dome size, plastochron and total number of leaf and bract and flower primordia that in fact are observed. However it comes about physiologically, the change in the primordia from being contributors to the growth of the apical dome to becoming competitors with it, is that it is all that is required in this model to result in the changes in partitioning that result from flowering, with its consequent reduction in primordium size at initiation and change in apical growth pattern.

The other model is a catastrophe model (Thornley and Cockshull 1980). By using equations from catastrophe theory it is possible to model the transition to flowering in *Chrysanthemum*. The basic assumptions are that:

1. the primordia produce an inhibitor of primordium initiation where inhibitor production is a function of primordium size on initiation;
2. the intrinsic growth rate of the system is constant; and
3. the inhibitor is diluted by the growth of the apex and is also degraded at a constant rate.

The switch to flowering is recognized by the production of small primordia – vegetative primordia can be up to 80% of the apex, but reproductive primordia may be just a few percent. By a judicious choice of starting values, after a time the model produces a sudden shift, of the size of the primordium at initiation, and of the plastochron, to much lower values, and at the same time the apical dome increases greatly in size. Implicit in this model is the assumption that the size of a primordium at initiation is a function of the amount of inhibitor produced by the new source. In other words, it is consistent with presuming that primordium area at initiation and inhibitor production are intimately related.

Both models rely on the assumption that the primordia produce an inhibitor of further primordium initiation and that a new primordium is initiated when the inhibitor level reaches a critical low threshold. They are both based on the operation of a diffusion–reaction system. Although primordium size relative to the apical dome may be specified as constant in the models, the absolute primordium size at initiation would essentially be determined by the plastochron, in the sense that the time of primordium initiation is specified by the inhibitor reaching a certain low level. In both these models a minimum size for a primordium is specified. In the first, this is a function of inhibitor concentration which has a specified low concentration when a new primordium will be formed, but the actual primordium is assumed to be a point. In the second, a minimum size for a primordium is specified. So in both it is really the plastochron that is being specified in terms of changes in concentrations of a hypothetical inhibitor in the apex. The challenge is to identify at primordium sites in the shoot apex cyclic changes in inhibitory substances or in gene expression which could be responsible for them.

8.4.4. *Changes in partitioning in the flower*

The formation of the flower is linked to striking changes in partitioning. There seems to be a step-wise reduction in relative primordium size ($2 \log_e r$) on formation of the floral organs (Table 8.2; Lyndon and Battey 1985). The first reduction (at the initiation of the sepals) may be linked to the increase in growth rate in the apical dome which typically occurs at the transition to flowering and which, in *Silene*, lasts only for one plastochron (Lyndon 1977b), so that the apex makes a spurt of growth between the initiation of successive primordia (the last leaf and the first sepal). This would be subsequent to any changes in partitioning, and reduction of primordium size, that there might have been at the initiation of bracts and which was modelled in *Chrysanthemum* (see above). There may be further reductions in primordium size at initiation of the petals or stamens. On the initiation of carpels in magnoliaceous flowers, the primordium size at initiation seems to remain essentially constant as the apical dome is used up (Lyndon 1997), suggesting that, at least in this instance, partitioning is determined by the size of the primordium at initiation and the availability of apical tissue.

So far, few measurements have been made, but since a change in phyllotaxis in the flower meristem is probably common to almost all flowers, we can infer that the reduction in primordium size when the floral organs are initiated is also common to almost all flowers. Even in those flowers in which the divergence angle does not change on flowering, the floral organs are initiated much faster than leaves and this would be consistent with a reduction in relative primordium size. This indeed is found in *Ranunculus* (Table 8.2), in which the petals are the only true

Table 8.2. *Reduction on size of primordia on flower initiation.*

Flower meristems: areas of primordia at initiation, relative to the apical cross-sectional area ($2 \log_e r$), and absolute size (area on apical surface)

Plant		$2 \log_e r$	Area on apical surface ($10^3 \mu m^2$)
Silene	leaves	0.71	7.5
	sepals	0.16	2.8
	stamens	0.11	1.4
Ranunculus	leaves	1.42	23.2
	sepals	0.34	7.8
	stamens	0.04	2.1
Impatiens	leaves	0.41	4.7
	petals	0.26	1.7
	reverted organs	0.20	1.6

Data from Lyndon and Battey (1985).

whorl. However, the sepal primordia have already shown the reduction in primordium size (Meicenheimer 1979).

A change in primordium arrangement on flowering is associated with a decrease in the sizes of primordia relative to the apical dome. This could come about either because of an absolute decrease in primordium size or because of an increase in apical dome size. This latter seems entirely plausible because one of the characteristic events of evocation is an increase in the size of the apical dome. *Humulus* and *Perilla* are the only plants so far examined in which the apex does not increase in size on flowering (Lyndon and Battey 1985). Probably in most plants there is both an increase in dome size and also a decrease in primordium size.

Although a concomitant of flowering, and perhaps a necessary condition, a reduced primordium size relative to dome size is not in itself sufficient to cause flowering. This is shown by reverted *Impatiens* in which primordium size relative to the apex was the same as in the flower, and yet the plants were forming leaves (Battey and Lyndon 1984). Similarly, when *Xanthium* was treated with gibberellic acid, the phyllotaxis was changed but the apices did not flower (Maksymowych, Cordero and Erickson 1976; see Section 8.4.1). The reduction in primordium size, at flower formation, therefore seems not to be causal to flower formation, but it may in some way be a necessary precondition for initiation and development of floral organs. Alternatively it is a necessary accompaniment of some other condition (e.g. reduced auxin content of the apex?) which is essential for flower formation.

8.5. Control of partitioning

Whether partitioning maintains the pattern of growth of the apex or whether it changes it, the question is raised: what determines the size of a primordium at initiation and what determines the size of the apical dome that remains? The question can also be posed another way: what determines the minimum distance from the apical summit that primordia arise?

What is likely to control primordium size at initiation? This has been modified experimentally by the application of auxin inhibitors to the apex (Meicenheimer 1981). Application of α-4-chlorophenoxyisobutyric acid (CPIB), an auxin antagonist, made a new primordium smaller; whereas N-1-naphthylphthalamic acid (NPA), an auxin transport inhibitor, made it larger. In both cases this resulted in a change of primordium arrangement (phyllotaxis) at the apex, because by changing the sizes of the newly initiated primordia, apical partitioning was altered. In both cases these changes in primordium size were brought about without altering apical growth rates.

These experiments, and others showing that application of auxins and auxin analogues can induce primordium formation and alter phyllotaxis, suggest that auxins are involved in determining primordium size at initiation. Young leaves and primordia are apparently sites of auxin synthesis (Lyndon 1994) and since the isolated apical dome requires auxin for growth and primordium formation, and so presumably does not have its own supply, the implication is that when leaves are initiated they begin *de novo* to synthesize auxin. The formation of leaf primordia may then be associated with the induction of sites of auxin synthesis at points on the apical dome where auxin concentration is lowest (Jackson and Lyndon 1990) (Fig. 8.8). The proposal is that the area of the apical surface over which auxin reaches a threshold concentration determines the size of the primordium, perhaps by determining the area over which plasticity of the surface layers is increased sufficiently for a primordial bulge to form. This would would also be consistent with a diffusion–reaction mechanism for determining primordium sequence or positioning.

An alternative view is that the size of a primordium is determined by biophysical factors, in particular the intrinsic wavelength of the tunica–corpus system (Green et al. 1996). What seems likely is that both the chemical and biophysical mechanisms exist and interact, each reinforcing the other. It would not be surprising if either system alone could, *in*

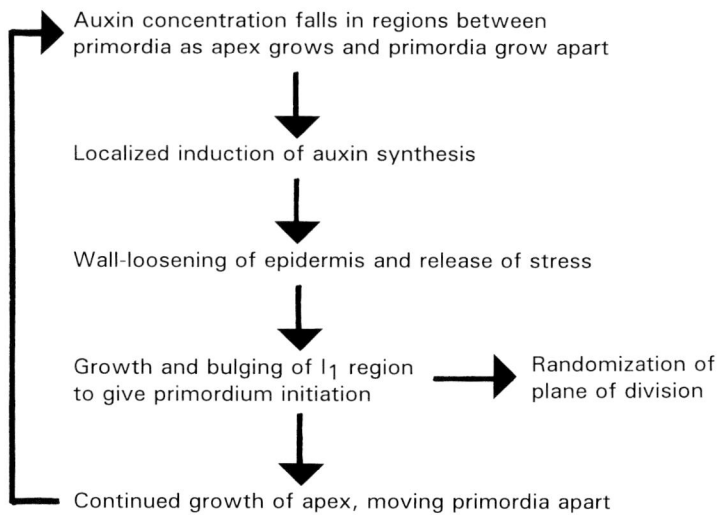

Auxin concentration falls in regions between
primordia as apex grows and primordia grow apart

↓

Localized induction of auxin synthesis

↓

Wall-loosening of epidermis and release of stress

↓

Growth and bulging of I₁ region → Randomization of
to give primordium initiation plane of division

↓

Continued growth of apex, moving primordia apart

Fig. 8.8 A possible model for the involvement of auxin in primordium initiation.

extremis, suffice for primordium initiation and the determination of primordium size, since plants often seem to have back-up or fail-safe systems so that they can maintain homeostasis in a hostile and rapidly varying environment.

The size of the apical dome in successive plastochrons relative to the newly initiated primordia depends on the size of the primordia at initiation and the distance from the apical summit that the primordia arise. What determines this minimum distance? The apical dome also still has to maintain its shape and structure. There must be some sort of mechanism for determining the maximum size of a primordium and for preventing primordia from arising on the apical dome, and so keeping it as a primordium-free zone. This may be simply because the growth rate is least on the apical dome, primordia necessarily forming where growth rate is fastest. Or there may be some sort of inhibitor of primordium initiation produced at the tip of the dome.

The size of primordia at initiation is at least partially under the influence of the apical dome. The width of a *Lupinus* primordium at initiation was estimated to be about 122° of arc on the apical surface (Snow and Snow 1952). When vertical radial cuts were made about 100° apart at presumptive primordium sites in the *Lupinus* apex, no leaves were formed at these sites, except in apices which were becoming floral or in apices in which the apical summit was split by a cut. When the apex was pricked the primordia formed were smaller and some occupied arcs of < 100°

(Snow and Snow 1955). Thus some influence from the apical dome seems to increase the minimum size of a primordium at initiation.

Whatever the mechanisms involved, changes in the growth gradient, or wall microstructure, or wall extensibility, are expected to be associated with changes in partitioning, and these should be worth looking for.

The Transition to Flowering

9.1. Phase change and developmental switching at the apex

The transition from vegetative to reproductive growth is a phase change from one stable state (vegetative) to another (floral) (McDaniel, Singer and Smith 1992). In monopodial inflorescences the flower meristems are formed only on side branches, except for the terminal flower, when one is formed. Below the inflorescence the later leaves are often successively smaller and bracts may be formed below the inflorescence and below each flower. In some inflorescences the bracts are vestigial or not formed at all so that the inflorescence axis bears only flowers on bare pedicels. In sympodial inflorescences the shoot apex itself first forms a flower and the growth of the inflorescence is continued from a bud or buds below the flower (Weberling 1992). Daisies and other similar composites at first appear to be terminal flowers but in fact are monopodial apices with the flowers borne laterally. Often composite 'flowers' are treated as flowers, with which they are wrongly compared, when in fact they are inflorescences.

The inflorescence axis often elongates more rapidly than the vegetative axis, and, in rosette plants and grasses particularly, most of the stem elongation is during the development of the inflorescence. Flower pedicels may also elongate rapidly. On the other hand, in the flower itself axis elongation is suppressed because the internodal cells necessary for elongation do not seem to be formed (Lyndon 1987a).

Flowers themselves characteristically consist of superimposed whorls of floral organs, usually not in the same arrangement as the leaves, and differentiated acropetally into sepals, petals, stamens and carpels, the latter normally occupying the summit of the apex so that meristematic activity there eventually ceases. Flower meristems are therefore determinate, whereas inflorescence meristems, such as racemes, may be indeterminate.

9.1.1. Competence to flower

The transition to flowering in most plants is probably influenced in some way by the environment in general (Lyndon 1992) but it has been studied mainly in plants in which flowering can be induced by photoperiod. This is so that the beginning of the flowering transition can be pinpointed by the experimenter, which is not possible in plants that are autonomous for flowering, i.e. in plants which do not depend on a specific environmental cue. In photoperiodically sensitive plants the flowering process can be divided into three stages (Evans 1969): (1) floral induction (as a result of photoperiodic action on the leaves), (2) evocation (the commitment of the shoot apex to flowering), and (3) realization, or floral morphogenesis (the actual formation of flowers). The ability of the apex to become switched to the flowering mode, i.e. to undergo phase change and become determined for flowering, depends first on its competence to respond to the floral stimulus.

What is the floral stimulus? What triggers the developmental switch from vegetative to floral? The existence of the floral stimulus is assumed because in photoperiodically sensitive plants induction has been shown to occur in the leaves, whereas it is the apex, some distance away, that reacts (Knott 1934). The signal passing from leaf to apex is usually assumed to be chemical in nature. Because a floral stimulus can be transmitted by grafting between plants of different genera (in the same family) and between different response types (short-day, long-day and day-neutral), the idea took hold that there was a universal stimulus common to all plants (Zeevaart 1976). Although the stimulus has been searched for, it has never been found. It has never actually been convincingly shown that the floral stimulus can be transmitted across family boundaries. Present evidence would be equally consistent with each family having its own floral stimulus. Some plants can be triggered to flower by growth substances. In long-day rosette plants, gibberellin applied to vegetative plants can make them flower in non-inductive short days. Gibberellins can also promote flowering (coning) in some conifers. In bromeliads, such as the pineapple, the floral stimulus seems to be auxins, auxin analogues or ethylene, all of which have been used commercially to promote flowering.

The floral stimulus can also be a negative stimulus, i.e. the removal of an inhibition of flowering. In *Ribes*, juvenility is apparently maintained by gibberellins produced in the roots, which inhibit flowering until the plants are large enough for some branches to be sufficiently removed from the influence of the roots (Schwabe and Al-Doori 1973). Explants of *Scrophularia* are prevented from flowering by cytokinins, which again seem to be produced by the roots (Miginiac 1972).

The floral stimulus produced by induced leaves could be required either to make an apex competent to respond to further stimulus, or to cause an already competent apex to become determined for flowering, or both. Competence to respond to a signal, and determination as a result, at the switch to flowering, are probably processes similar to those at other developmental switches in the plant. Plant growth substances are effective stimuli for the acquirement of competence and for determination in callus; and for determination in vascular element differentiation (Sachs 1981, Jackson and Lyndon 1990). It seems likely that the floral stimulus may take the form of an increase in some components and a decrease in others (Bernier 1988), and perhaps only transiently, or long enough for the phase change or determination to be accomplished. The floral stimulus is probably no different from other stimuli that might be required to trigger determination in other developmental processes.

9.1.2. *Vernalization*

Some plants require vernalization to flower, it being an exposure to a cold period which makes the plant then capable of subsequently flowering when exposed to a warmer, favourable environment, often long days (Vince-Prue 1975, Bernier, Kinet and Sachs 1981). It is the shoot apex that perceives the cold, as shown by experiments in which either the shoot tip or the rest of the plant was exposed to the cooling treatment, or in which only the shoot apical bud was cooled by cooling coils (Vince-Prue 1975). Since it is the shoot apex that eventually responds, by flowering, the presumption is that vernalization in some way alters the cells of the shoot apex so that they become competent to respond to further stimuli (Jackson and Lyndon 1990). Vernalization produces no detectable developmental changes to the shoot apex except for possibly (in wheat) a reduction of leaf primordium size relative to apex size (Griffiths, Lyndon and Bennett 1985). The changes caused by vernalization are presumably at the molecular level and cause a change in competence (but not determination) of the apex, so that it has now become capable of responding to further signals (e.g. long days) and to make the switch to flowering. When plants requiring vernalization were treated instead with the DNA demethylating agent 5-azacytidine, non-vernalized plants flowered significantly earlier than the untreated controls (Burn et al. 1993). Cold- or 5-azacytidine-treated plants also had reduced levels of 5-methylcytosine compared with non-vernalized controls. It was suggested that vernalization causes desuppression of the activity of a shoot apex-specific gene in the gibberellin synthesis pathway.

Vernalization cannot be substituted by gibberellin in cereals. Genes for a range of vernalization response have been introduced into wheat (Law, Worland and Giorgi 1976) but have not yet been examined at the molecular level to see what the gene products are, or what they might do.

9.1.3.   Phase change at the cellular level

Underlying the switch from vegetative to flowering growth is a change in the properties of the apical cells themselves. When a vegetative shoot apex is grown in culture it continues to produce leaves. An isolated flower meristem makes a flower. Chailakhyan et al. (1974) showed that tobacco callus derived from vegetative plants continued to make vegetative structures whether grown on media with or without glucose. Callus derived from flowering plants grew vegetatively if grown on media deficient in glucose, but with glucose, flowers formed (Fig. 9.1). Therefore there were two kinds of vegetative callus: that which could not form flowers whatever the culture medium, and that which was potentially floral in the presence of glucose. The first was determined for vegetative growth, the

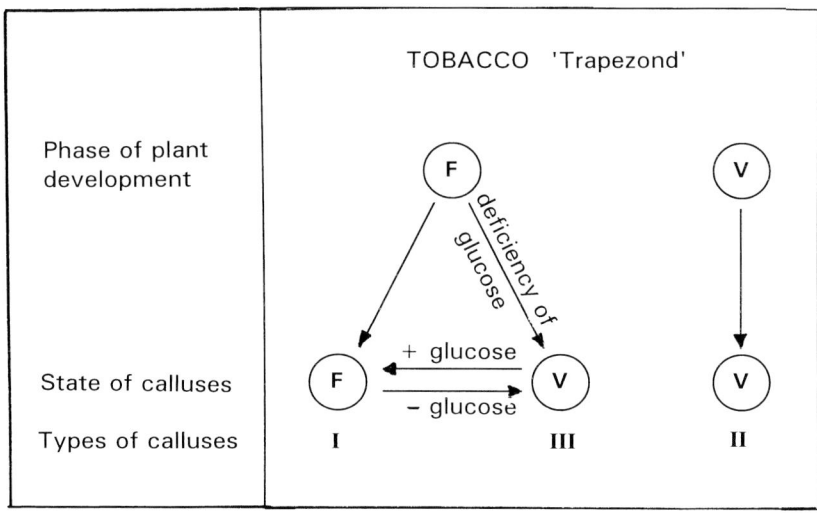

Figure 9.1 Callus model of flowering. Callus from vegetative (V) tobacco plants (*Nicotiana tabacum*, cv. Trapezond) regenerates only vegetative (V) plants. Callus from flowering plants (F) regenerates vegetative plants (V) in the absence of glucose in the medium but flowering plants (F) when glucose is present. From Chailakhyan et al. (1974).

second was competent to respond to glucose and to become determined for floral growth.

Determination is an operational criterion that can be shown only by experiment. Cells are determined when they have the same developmental fate whether left *in situ*, isolated from the rest of the organism, subjected to experimental treatments or transplanted to a different position in the organism (McDaniel 1984). Apices are determined for a particular developmental state once they cannot be diverted to an alternative pathway of development. If they are in one determined state but have the potential to be switched to another, they are said to be competent to undergo phase change. Phase change from one determined state to another is well illustrated by plants showing juvenile and mature forms. Competence is shown when cells exposed to a signal or treatment respond by developing along the new, and expected, pathway.

Ivy is the classic case in which the vegetative form shows different morphological characteristics from the flowering form. The shoot apices of the two forms are also distinguishable from each other, and the cells are larger in the juvenile apex (Table 9.1). The persistence of juvenile and adult characteristics in cells in culture (Stoutemyer and Britt 1965, Robbins and Hervey 1970, Polito and Alliata 1981) and the experiments

Table 9.1. *Characteristics of shoot apices of juvenile and adult ivy* (Hedera).

	Juvenile	Adult
Mode of growth	vegetative	flowering
Phyllotaxis	distichous	spiral
Leaf form	lobed	entire
Plastochron (days)	4.2	3.2
Mean apex width (μm)	137	294
Mean cell height (μm)	11	9
Mean cell width (μm)	10	8
Approximate mean cell volume (μm^3)	1100	596

Data from Hackett, Cordero and Srinivasan (1987) and Stein and Fosket (1969).

with tobacco callus show that the phase change from vegetative to floral resides at the cellular level and does not depend on organization at a higher level although it may be responsible for it. Once changed to the floral state, the apex usually remains in this state but in some plants the apex can revert to vegetative growth. Thus, although the phase change is from one stable state to another, and normally only in one direction, it can sometimes be reversed. The mature form of ivy can be made to revert to the juvenile form if it is treated with gibberellin (Rogler and Hackett 1975). Reversion from the mature to the juvenile phase can also be brought about in callus by treatment with gibberellin (Miller and Goodin 1976). Mature callus treated with gibberellin slowly reverts over a period of weeks to the juvenile form as shown by the characteristics of the plants that are regenerated from it.

Changes in competence and determination in the tobacco shoot apex have been examined, especially by McDaniel and colleagues. The tobacco strain *Nicotiana sylvestris* is a long-day plant and, when grown in inductive conditions, flowers only after about 36 leaves have been formed. If a plant is decapitated then an axillary bud will grow out, taking over the function of the main shoot, producing leaves, and only after it forms $\geqslant 36$ leaves does it flower (Dennin and McDaniel 1985). Similarly, if a lower axillary bud is excised and transplanted so that it grows on independently, it will also make leaves to a total of about 36 and then flower. At excision the apices of such buds are therefore determined for vegetative growth. If a bud is excised from near the apex of a plant that is beginning to flower, this bud when excised makes only about 10–16 leaves and then flowers. It is therefore considered determined for flowering. Using this test, it was shown that the topmost three to four buds on a

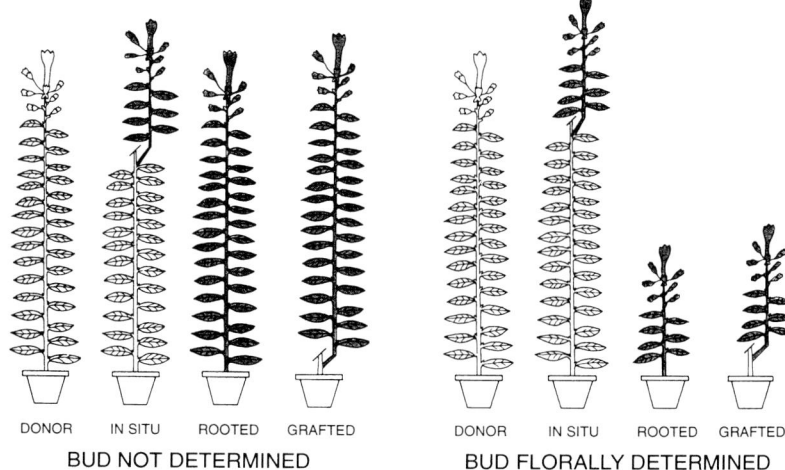

Figure 9.2 Determination of buds on tobacco plants. An axillary bud that grows out from a decapitated plant forms about the same number of leaves that the intact apex would have made. When an axillary bud is excised and rooted, or grafted to the base of a decapitated plant, if it produces the complete complement of leaves for the whole plant, then the bud was not determined for flowering when it was excised. If it forms only a few leaves before making flowers, then it was already determined for flowering at the time of excision. From McDaniel (1996).

flowering plant were determined for flowering but that buds below this were determined as vegetative. Determination for flowering therefore takes place after about 33 leaves have been initiated. Similar conclusions were drawn from experiments with the day-neutral strain of tobacco *N. tabacum* cv. Wisconsin 38 (Fig. 9.2) (McDaniel et al. 1985). These experiments demonstrate when the developmental switching occurs in the shoot apex of plants autonomous for flowering. The cellular changes involved can be more easily followed in plants that require a signal to flower so that the onset of flowering can be pinpointed experimentally, and the processes of commitment of the apex to flowering – evocation – can be followed.

9.2. The transition to flowering at the shoot apex: evocation

Evocation consists of the processes that occur at the shoot apex during determination for flowering. In plants that require a specific environmental stimulus to flower, the precise moment at which the transition of the apex to flowering begins can be defined experimentally. The beginning of evocation is taken to be when the floral stimulus reaches the apex. The end of evocation is when inhibitors applied to the apex can no longer prevent it becoming floral rather than continuing vegetative growth (the 'point of no return') or, if it cannot be pinpointed experimentally, the time at which flower morphogenesis begins (Bernier 1988).

The visible, morphological events of evocation in most plants are (Bernier 1988):

1. enlargement of the apical dome;
2. reduction in the sizes of primordia at initiation relative to the apical dome;
3. shortening of the plastochron; and
4. precocious development of the youngest axillary buds, indicating a release from apical dominance.

9.2.1. *The cellular events of evocation*

At the cellular level in the shoot apex, evocation is characterized by:

1. increase in respiratory activity;
2. increase in RNA, protein synthesis and accumulation;
3. changes in protein complement;
4. increase in rate of growth and cell division;
5. synchronization of cell division; and
6. commitment of the apex to flowering.

These events of evocation may occur more or less in this order or may be more or less simultaneous (Bernier 1971, 1979, 1984, 1988, Bernier, Kinet and Sachs 1981, Lyndon and Francis 1984, Lyndon and Battey 1985) and have been documented in most detail for relatively few plants (*Sinapis*, Bernier, Kinet and Bronchart 1967, Kinet et al. 1971, Bernier et al. 1974, Havelange, Bernier and Jacqmard 1974, Havelange and Bernier 1983, Havelange, Bodson and Bernier 1986; *Silene*, Miller and Lyndon 1976, 1977; *Pharbitis*, Herbert, Francis and Ormrod 1992; *Xanthium*, Jacqmard et al. 1976, Havelange 1980, *Lolium*, Evans, Knox and Rijven 1970, Jacqmard et al. 1993; *Chenopodium*, Seidlová 1980, Opatrná et al. 1982, Seidlová and Sádlíková 1983). The characteristic shortening of the cell cycle and the transient increase in growth rate are discussed further in Chapters 3 and 4.

9.3.   Changes in respiratory activity

Changes in respiratory metabolism are indicated by changes in respiratory enzyme activity. These include increases in glucose 6-phosphate dehydrogenase activity (a marker for the pentose phosphate pathway) (Auderset et al. 1980, 1985), and in glyceraldehyde 3-phosphate dehydrogenase (Orr 1987) and succinoxidase activities (Jacqmard 1978) (markers for the glycolytic pathway). There is also a corresponding increase in the number of mitochondria in the apex (Havelange, Bernier and Jacqmard 1974, Huang et al. 1994) and, in *Sinapis*, a fragmentation of the vacuoles. Both of these events involve increased membrane synthesis. Similar changes in *Nicotiana* are found first in the tunica (epidermis) and a little later in the corpus cells (Kanchanapoom and Thomas 1987a, b). Respiratory changes are probably linked to the changes in starch amount and distribution within the shoot apex on transition to flowering (Havelange, Bernier and Jacqmard 1974, Havelange 1980, Seidlová 1977), which, in turn, are probably related to the supply of assimilates such as sucrose, which increases in the apex on flowering (Bodson 1984, Bodson and Outlaw 1985).

9.4.   Changes in protein complement

On the basis that changes in gene activity should be evident as changes in the types of proteins synthesized, we would then expect changes in protein synthesis and composition in the shoot apex during the transition to flowering. On floral induction in *Silene*, the concentration of protein in the shoot meristem cells increased from the fourth day of induction onwards (Miller 1976, Taylor et al. 1990). A transient rise in protein concentration was also found only 2h after the beginning of induction. None of these changes in protein concentration during the 7 long days of induction were accompanied by qualitative changes in the protein complement (Taylor et al. 1990) and therefore reflect changes in the growth rate of the cells and the rate of protein accumulation rather than qualitative changes in gene expression.

Early attempts to show changes in protein composition of the shoot apex on floral induction were limited by the techniques available and only very minor or no changes could be found (Marushige and Marushige 1962, Steward et al. 1971, Miller 1976). With more sensitive detection techniques with two-dimensional polyacrylamide gel electrophoresis (2-D

PAGE), far better protein separations became possible. However, the differences in total protein composition detected have only been well into the flowering transition and long after evocation. In shoot apices of *Silene*, carefully dissected out so that the tissue samples consisted only of the apical dome plus the youngest pair of leaf primordia, no changes could be detected in the protein composition of the apices during the first 5 long days of floral induction and evocation. Changes were detected only 8 days after the beginning of induction, when evocation was complete (7 long days give 100% flowering) (Francis, Rembur and Nougarède 1988, Taylor et al. 1990). In the evoked apices at the end of evocation, on day 8, 107 new proteins were found and 182 others were identical to those found in the uninduced plants (in short days); 17 proteins unique to the uninduced apices disappeared (Francis, Rembur and Nougarède 1988). Similar results were obtained in confirmatory experiments (Taylor et al. 1990). Similarly, in *Sinapis*, careful analysis showed differences in protein composition only 10 days after the beginning of induction, when small flower buds had formed (Cremer, Van de Walle and Bernier 1985). The differences affected about 10% of the protein complement and most of the new spots appearing in the floral apices could be attributed to specific floral organs (Cremer 1992).

Similar results were obtained for *Chrysanthemum*. Ten new polypeptides appeared in the prefloral meristem and two (unique to the shoot apex) disappeared (Rembur and Nougarède 1989). In reproductive meristems, four new spots appeared and two disappeared (one in both vegetative and prefloral meristems, the other specific to prefloral meristems). Of the > 500 spots on the gels these changes involved $< 2\%$ of the proteins. The main changes occurred at the transition from vegetative to prefloral (i.e. bract initiation) growth at the apex. Analysis by 2-D PAGE therefore shows relatively few early differences in the apical meristem itself. The differences found are fairly late in the floral transition and can probably be ascribed to the new proteins specific to the various floral organs.

Labelling would be expected to show differences more sensitively and therefore earlier. In the *Sinapis* shoot apex, evocation becomes irreversible after 44h from the beginning of induction by a long day (Kinet et al. 1971). A comparison of the fluorographs of 2-D PAGE gels of extracts of apical meristems of vegetative plants (kept in non-inductive short days) and evoked plants supplied with [^{35}S]methionine for 2h, at 50h after the beginning of induction and about 6h after the completion of evocation, showed that six proteins (out of 400–500 detected) were synthesized less in the evoked meristems and 14 proteins were synthesized relatively more than the rest in the evoked meristems; also 16 new proteins were synthesized in *Sinapis* meristems at evocation, as shown by the incorporation of [^{35}S]methionine into apical dome proteins (Lyndon, Jacqmard and Bernier 1983). These changes represent very small changes in the total protein composition. Presumably the changes are not only in relatively few crucial proteins, but also only in those cells of the apical dome that are about to make new primordia or produce the floral or inflorescence meristem. In *Sinapis* the use of labelling allowed the detection of differences after 54h, at the end of evocation, whereas by staining differences could not be detected until 5 days (120h). When these labelling experiments with *Sinapis alba* were repeated in more detail, only one new protein could be detected before the formation of floral organs, but even then fewer than 10 new proteins could be detected in the apex itself (Cremer, Van de Walle and Bernier 1992).

When similar experiments were done with *Pharbitis*, about 1000 polypeptides were detected on 2-D gels either by silver staining or by autoradiography after labelling of the apices for 2h with [^{35}S]methionine (Araki and Komeda 1990). But only five polypeptides were different between buds of plants evoked to flower by being given 1 short-day, and the buds of vegetative plants prevented from flowering by being given a short burst of light in the middle of the night (known as nightbreak treatment or NB) after the short-day. Two of these polypeptides increased in short-days but remained low in NB, a third increased in short-day then decreased (but showed no change in the labelling experiment), and the fourth and fifth decreased in NB. These differences were retained or magnified with time from days 1 to 16. With a slightly different approach, but also with *Pharbitis*, the translation products of mRNA of induced and non-induced apices were examined, but with very similar results (Ono et al. 1988). Although about 400 polypeptides could be distinguished on the gels, there were only three present in the induced but not in the non-induced apices, even though the inductive treatment was greater (given on three successive cycles, i.e. 3 short-days versus 3NB). There were minimal differences between the leaf, cotyledon, petiole, hypocotyl and root (all, of course, 'vegetative' organs), although there were at least six organ-specific polypeptides. However, the main message from all these experiments is that remarkably few changes can be detected in the protein complement or in the proteins synthesized as a first result of evocation at the shoot apex.

The relatively few changes in polypeptides revealed by PAGE must reflect, in part, the limited resolution of the technique. When the number of genes active in an individual cell is compared with the maximum of about 1000 polypeptides detected on 2-D gels, it can be seen that where differences in polypeptide profiles emerge they are probably underestimates. Where no differences are resolved they may still mask subtle differences at the mRNA level. mRNA extractions coupled with the polymerase chain reaction (PCR) should, in future, enable closer scrutiny of changes in the protein complement during evocation. However, care must be taken to distinguish qualitative changes that commit the apex to a floral mode of growth from those changes that are linked to growth. Model photoperiodic systems, where treatments exist which are transient photoperiodic perturbations of otherwise inductive conditions (e.g. the NB in *Pharbitis*, the dark-interruption of long days in *Silene*), will be critical in this regard. In *Silene* the possibility was raised that the changes may have been related to the synchronization of cell division, which occurs on days 7–8 after the beginning of induction, rather than to flowering *per se* (Francis, Rembur and Nougarède 1988). The synchronous cell divisions in the apex could be suppressed by inducing the plants by 7 long-days, then placing them in darkness for 48h (see Fig. 4.5), a treatment which delayed, but did not prevent, flowering (Grose and Lyndon 1984). However, the same changes in protein complement were found as in induced apices on day 8, showing that the protein changes were indeed intrinsic to flowering and not concerned with division synchronization (Taylor et al. 1990). In future, *in situ* hybridization in these systems should resolve tissue-specific changes at the shoot meristem from those that are linked to cell division *per se*.

Changes in protein composition on evocation have also been shown by immunological techniques, but these are very much less sensitive to numbers and amounts of proteins. Immunological techniques have shown changes in protein composition of the *Sinapis* shoot meristem as a result of induction and evocation. An antigenic protein (A) in vegetative buds, and characteristic of young leaves > 3mm long (Pierard, Jacqmard and Bernier 1979), disappeared in induced plants 96–240h after the start of induction, when the flowers began to be formed. A second and third antigen (B and C) began to increase in concentration in induced buds 30 and 96h respectively after the start of induction and continued to increase as the flower buds developed up to 240h. However, neither of these antigens was specific to the induced state since they were also detected, but at much lower concentration, in vegetative buds (Pierard,

Jacqmard and Bernier 1977). Proteins B and C were also found in the young leaves of the vegetative shoot, but after induction they appeared in the sepals and pistil although absent from the petals and stamens (Pierard, Jacqmard and Bernier 1979).

The occurrence of these three antigens was looked at in more detail using immunofluorescent tagging of the antibodies to locate and measure these antigens in sections of *Sinapis* apices (Pierard et al. 1980). Protein A was uniformly distributed throughout the vegetative apical meristem, in central and peripheral zones. At the end of evocation (48 and 96h) its concentration increased, but by 240h it had disappeared from the floral meristem. Proteins B and C were localized together in the sections and were shown to begin accumulating 36h after the beginning of induction, that is 8h before the apex becomes irrevocably committed to flowering (Kinet et al. 1971) thus showing that changes to the protein composition, characteristic of the flower, begin during evocation. As flower bud formation progressed, these proteins increased in concentration. Proteins B and C were also uniformly distributed throughout the evoked meristem. Since not all meristems showed fluorescence at 36 and 40h, and 44h is the point after which flowering cannot be prevented by application of inhibitors to the apex (Kinet et al. 1971), this would be consistent with B and C beginning to increase in concentration as a result of the commitment of the apex to flowering. Proteins B and C, as measured by immunofluorescence, were not detectable in the vegetative meristem (Pierard et al. 1980). It was also noted that the appearance of B and C was relatively late compared with that period of evocation (up to 24h) which was most sensitive to inhibitors of RNA synthesis (Kinet et al. 1971). Again, this is consistent with these proteins being the result of evocation rather than part of it.

Further investigation was carried on by raising antisera to the individual floral organs to see if floral-specific antigens could be detected earlier, at the transition to flowering (Jacqmard, Lyndon and Salmon 1984). Three proteins specific to the flowering state were detected by immunodiffusion and immunoelectrophoresis, two stamen-specific and one pistil-specific. There were no antigens specific to sepal or petal. The localization of these antigens was examined by immunofluorescence in longitudinal sections of young developing flowers. The stamen antigens were exclusive to the pollen grains, especially the exine, and the pistil antigen was exclusive to the stigma and the style. None of these flower-specific antigens appeared before the flowers actually formed, and are therefore different proteins from those that increased in the meristem at evocation. These flower proteins are probably concerned with the male–female recognition system.

The small number of proteins detected by immunological techniques suggests that they are probably some of the most abundant proteins in the plant extracts. The proteins that disappear from the apex on induction may well be the subunits of Rubisco (the enzyme ribulose 1,5-bisphosphate carboxylase, the principal enzyme of photosynthesis), but this has not been demonstrated. If the proteins characteristic of evocation are in relatively low concentration then these immunological techniques may be quite inadequate to detect them. The immunological techniques also have the drawback that they depend on proteins being antigenically active, so the rabbits used to produce the antibodies 'select' the protein for investigator, a selection that should be under the control of the biologist.

9.5. Changes in gene expression

Changes in gene expression during the transition to flowering have now been shown by cDNA cloning technology. In *Sinapis alba* a group of six cDNA transcripts was identified. The transcripts were expressed only at low concentration in vegetative apices but accumulated to a maximum 2–10 days after the beginning of induction (Melzer, Majewski and Apel 1990). These transcripts did not seem to be involved in cell division as such because, although they were more prevalent on the flanks of the apex, they were in very low concentration at the summit of the floral apex, in which cell division is also rapid in *Sinapis* (Bodson 1975). Further work has shown that two genes, *SaMADS A* and *SaMADS B*, isolated because they have MADS-box sequences, are up-regulated in *Sinapis* shoot apices 48 and 72h respectively after the beginning of long-day induction (Menzel, Apel and Melzer 1996). This is 2 and 5 days before the expression of the floral homeotic genes *AP1* and *AG* (see Chapter 10). In tobacco (cv. Samsun; a day-neutral plant) a cDNA clone (FA2), transcripts of which were present in only very small amounts in the mature vegetative apex, increased in the shoot apical meristem during the transition to flowering and was highly expressed in the developing petal, stamen and pistil (Kelly et al. 1990). In tomato, activation of the meristem, as shown by increased gene expression and enzyme activity, was in the cells at the base of the meristem rather than in the central zone (Fleming and Kuhlemeier 1994).

Changes in gene expression were followed in cultures of thin cell layer explants (TCLE) of cv. Samsun tobacco (Meeks-Wagner et al. 1989). Six different gene families (FB-1 to FB 7-6) were identified. FB7 transcripts were poorly expressed in TCLE on the vegetative programme (in which flowers formed only late in culture) but had increased by day 7 in TCLE given kinetin, which induced early flower formation. This transcript then decreased but increased again by days 23–25 when the flowers were forming. However, none of the transcripts was floral meristem-specific. This is perhaps not too surprising since the whole process of TCLE is designed to elicit only certain programmes from TCLE that are competent for all (Tran Thanh Van 1981). This in itself might indicate that the changes in gene expression in different organ types are quantitative and that qualitative changes are only associated with changes in competence of cells during developmental switching. Very few changes, in either gene expression or protein complement, have been found at the transition to flowering. This points to the commitment of the apex and to the events of evocation depending on changes in expression of only very few key genes. It remains a challenge to find out what these genes are and what aspects of cell functioning they control.

9.6. The cell cycle and evocation

The events of evocation in *Sinapis* are closely tied to a synchronous cell cycle, which occupies the whole of evocation (see Fig. 4.4) (Bernier, Kinet and Bronchart 1967, Francis and Lyndon 1985). In *Sinapis* it has not been possible to dissociate the synchronous cell cycle from flowering by treatments which altered the timing of the first peak of cell division (Bernier et al. 1974).

In *Silene* there is also a synchronization of cell division (see Section 4.4.1), but this does not take place until the end of evocation, 8 days after the beginning of induction, and just before the formation of the first flower (Francis and Lyndon 1979). Because synchronization also occurs just before the formation of third-order flowers, 3 weeks after induction and the return of the plants to non-inductive conditions, it seems to be a precursor of flower formation rather than an event of evocation *per se* (Lyndon 1987b). Synchronization of cell division in *Silene* can be prevented by placing the plants in darkness for 48h immediately after induction, but flowering is not prevented, only delayed by 48h (Grose and Lyndon 1984) (see Fig. 4.5). This shows that synchronization is not

Table 9.2. *Changes in the molecular exclusion limit in shoot apices of* Silene coeli-rosa *on the transition to flowering.*

Probe size (daltons)	559	536	665	749	874	2268
Non-induced apices	3/3	11/15	2/6	0/6	0/1	0/1
Induced apices	3/3	2/14	0/8	–	–	0/1

Apices were injected with fluorescent oligopeptide dye conjugates of various molecular sizes and the proportion of trials in which the dye moved freely and rapidly throughout the apex was recorded. Apices induced to flower were sampled at 15:00 hours on the eighth day after the beginning of induction when synchronous cell division was taking place (see Fig. 4.5). Induced apices show a reduction in the freedom of movement of the conjugate probes, indicating a reduction in the molecular exclusion limit from molecules about 600 daltons to molecules about 500 daltons.
Data from Goodwin and Lyndon (1983).

essential for flowering in *Silene* and that, when it occurs, it seems more linked to the onset of flower morphogenesis than to evocation itself (Lyndon 1987b). The synchronous divisions are concentrated in the peripheral zone of the apex, but are not obviously clustered at the primordium sites and show the same sort of distribution as in the vegetative apex (Francis 1992).

Changes in the molecular exclusion limit in the cells of the apex in *Silene* appear to be linked to synchronization and similarly may normally accompany flowering without being integral to it. In prefloral apices of induced plants, sampled on the eighth day after the beginning of induction, when there is a peak of mitotic index (see Fig. 4.5), the molecular exclusion limit reduced slightly, indicating that the apical cells became more isolated from each other (Table 9.2) (Goodwin and Lyndon 1983). This was the opposite of what was anticipated; it was thought that when the cells became synchronous this might have been because of a reduction of diffusion barriers between the cells, not an increased restriction. When synchronization in *Silene* was eliminated by 48-h darkness after floral induction, flowering was delayed but not prevented, but the change in the molecular exclusion limit was also eliminated (Santiago and Goodwin 1988). Changes in the molecular exclusion limit therefore seem to be linked to changes in the cell cycle and not to flowering itself.

Changes in the cell cycle in *Silene* that do seem to be closely linked to evocation are those occurring at the very beginning of induction and are triggered by red or far-red light (see Section 4.4.2). As soon as *Silene* plants are transferred from non-inductive short-days into continuous

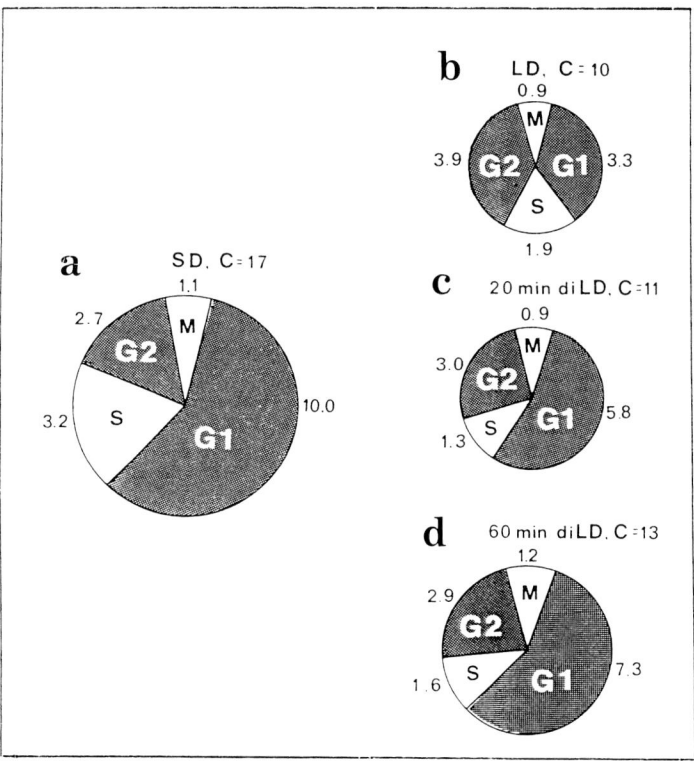

Figure 9.3 Effect of a dark interruption on the cell cycle in the shoot apex of *Silene*. The cell cycle (C) is reduced from (a) 17h in short-days (SD) to (b) 10h after 1 long-day (LD). With a dark interruption at the beginning of the long day (diLD) of 20 (c) or 60 min (d), the cell cycle still shortens, but the relative lengths of G_1 and G_2 remain the same as in SD. From Ormrod and Francis (1987).

light, the cell cycle shortens immediately and the rate of DNA synthesis doubles (Francis and Lyndon 1978a, b, Ormrod and Francis 1986b). The ratio of cells in G_2/G_1 characteristically increases (Francis 1981a) because of a marked shortening of G_1 (Ormrod and Francis 1987). This shortening of the cell cycle, from 17–19 to 10–13h, lasts for only a single cell cycle. Flowering is >90% inhibited by interposing only 20 min of darkness at the beginning of the continuous-light inductive treatment (Ormrod and Francis 1986a). This short dark period does not prevent the shortening of the cell cycle (Fig. 9.3) but it does prevent the increase in the G_2/G_1 ratio, which therefore seems in some way to be linked to evocation, in the same way that the synchronous cell cycle is in *Sinapis*. A very early increase in mitotic index (Jacqmard et al. 1993) and the much later increase in the G_2/G_1 ratio in the shoot apex were also correlated with flowering in *Lolium* (Gonthier and Francis 1989).

In *Sinapis*, *Silene* and *Lolium* there are therefore cell cycle changes in the shoot apex which seem to be integral to the transition to flowering, although the changes are different in at least *Sinapis* and *Silene*. In *Sinapis* especially, many of the cellular changes found during evocation could be consequences of the cell cycle changes rather than themselves being causal to flowering. This then focuses attention on the cell cycle changes, especially those occurring at the beginning of induction. How might these be causal to flowering? Are they perhaps necessary for switches in gene expression, or do they in turn result from changes in expression of flowering genes?

9.7. What are the significant changes in evocation and flower initiation and what do they accomplish?

Many of the changes occurring during evocation are similar to those found when growth rate increases and the cell cycle shortens. This raises the question whether the events of evocation are simply concomitants of an increased growth rate that is characteristic of the apex in transition to flowering. In axillary buds released from apical dominance by cytokinin, there were increases in respiratory activity, cell division rate, primordium initiation rate, RNA and DNA synthesis, and changes in starch accumulation, and yet these buds were strictly vegetative, in transition only from slow to rapid growth (Usciati, Codaccioni and Guern 1972).

In order to try and distinguish any events essential for flowering from those simply associated with a changing growth rate, attempts have been made to dissociate the two processes. By growing *Silene* plants at lower temperature, the growth rate of the apex was reduced but the plants flowered even more readily (Miller and Lyndon 1977). *Silene* apices treated with gibberellic acid increased their content of RNA considerably but this had no effect on flowering (Miller and Lyndon 1977); however, this could have been because the extra RNA was not directly involved in the floral transition. The small changes in protein complement that have been detected (see above) have not unequivocally been shown to be linked to flowering rather than with the simultaneous changes in growth rates in the apex. So far, specific links between biochemical and enzymic changes in the apex and evocation have not been shown. Only the cell cycle seems to be more critically linked to evocation.

Evocation is determination of the shoot apex for flowering. It seems to be determination for suppression of the expression of genes such as *EMF* (embryonic flower) in *Arabidopsis* (Sung et al. 1992), and the activation of genes which switch the apex to a flowering mode, such as *LEAFY* in *Arabidopsis* and *FLO* in *Antirrhinum* (Bradley et al. 1993). Whether the activation of the homeotic genes determining the identity of the different floral organs (see Chapter 10) follows evocation or is partly concurrent with it has not yet been studied. Since a plant such as *Silene* forms a pair of leaves before each flower in the inflorescence, but does not revert to vegetative growth, the homeotic flowering genes are presumably activated at the onset of the formation of each individual flower which, in *Silene*, coincides with the synchronization of cell division in the meristem (see above).

Although evocation may appear to be a unique developmental switch because it is concerned specifically with the transition to reproductive growth and also requires a special trigger, the floral stimulus, it is probably not essentially different from any other developmental switch. The floral stimulus is probably no different from other stimuli that may be required for determination in other developmental processes. Stimuli are required for determination in callus (Lyndon 1990) and for determination in vascular element differentiation (Sachs 1981), and are the common plant growth substances. Also in flowering, the stimulus can be a plant growth substance (e.g. auxin or ethylene in the pineapple, gibberellin in long-day rosette plants). Evocation is probably just another example of a change in gene expression, but with an end result – flowering – that, to the observer, is more spectacular than most.

10

The New Floral Meristem

10.1. New developmental patterns at the flowering apex

When a plant flowers, gene expression in the plant reaches its fullest extent. Each type of organ within the flower probably expresses about 25 000 genes. In the anther and ovary there are about 10 000 genes not expressed elsewhere in the plant. The petal, on the other hand, has about 7000 genes expressed in common with the leaf, but with no other plant organs, but also has distinctive gene expression of its own (Goldberg 1986).

The shoot apex changes dramatically in the way it functions, as activation of new gene expression causes switching into new developmental pathways and the formation of the different floral organs. Many cell types are specific to the flower. Organs, particularly the stamen and carpel, are unlike those found anywhere else in the plant. The formation of the reproductive structures and gametes is unique to the stamen and carpel. Meiosis is integral to gamete production and occurs only in the anthers and ovary, so there are important changes in the nature of cell division. Also, cytokinesis may be temporarily suspended during gametogenesis, and cell division in the anthers becomes synchronized (Bennett 1976).

The shoot apex itself may become a flower (as in cymose inflorescences) but more often the apex continues functioning in much the same way as it does in the vegetative mode, at least for some time, and it is the shoot apices of the axillary buds or their branches that become the first flowers. Because of the new patterns of gene expression that are seen as the flower develops, the flowering shoot apex has become the focus of much recent research into the molecular and genetic control of plant development.

The morphological changes at the formation of the inflorescence are essentially changes in branching pattern of the axis, often with precocious development of axillary buds. There are more specific changes at the onset of the formation of the flowers themselves. The first indication of the formation of a flower is often a change in the arrangement of the primordia at the flower meristem (see Chapter 7). If, as has been argued (Lyndon 1990), all that would be required to bring about a change in patterning at the shoot apex might be a change in the nature of the diffusing molecules in a morphogen field, then these first stages in flower formation could involve changes in the expression of no more than three or four genes (perhaps more for the suppression of internodes). *FLORICAULA* in *Antirrhinum* and *LEAFY* in *Arabidopsis* are homeotic genes that are required for the disappearance of internodes in the flowers (Bradley et al. 1993). The *FLORICAULA* and *LEAFY* genes are required for flower formation to proceed; when these genes are mutant a vegetative 'inflorescence' forms, internodes are retained and flower formation does not proceed. One of the other changes required to cause the first steps of flowering would be activation of a gene causing a change in a morphogen responsible for primordium size. In *Antirrhinum*, the *SQUAMOSA* gene suppresses the formation of bracts and bract-like leaves (Huijser et al. 1992) and, in conjunction with *FLORICAULA*, apparently regulates primordium size (Carpenter et al. 1995). Another change would be activation of a gene for altering branching pattern and elongation of the inflorescence axis.

In families thought to be less advanced evolutionarily, flower formation may not be associated with change in primordium arrangement and patterning. In the Magnoliidae the sepals, petals, stamens and carpels all may continue the helical sequence of primordia shown by the leaf primordia, with a sustained divergence of 137.5° through the leaves and the floral organs (Erbar 1988). In some genera only the perianth is initiated as a whorl, as in *Ranunculus* (Meicenheimer 1979). However, in all these flowers internodes disappear and floral organ primordia are reduced in size relative to the shoot apical dome, so these are perhaps the critical changes on transition to formation of flowers. A reduction in primordium size may be obligatory for flowering, leading normally, but not necessarily, to a change in primordium arrangement. Similarly, although primordia reduced in size normally develop as floral organs, they do not always do so, as in reverted *Impatiens*, in which they form leaves (Battey and Lyndon 1984).

The observation that the stamens in some Magnoliales (especially the later stamens) do not arise in the temporal sequence corresponding to their spatial sequence (Erbar 1988), suggests a prepatterning which is then realized by later development. Also of interest in this respect is the initiation of primordia across the apex, irrespective of floral whorl, which is found in some Leguminosae. The spatial sequence is helical but the temporal sequence bears no relation to this; it sweeps from the front to the back of the floral apex. Conversely, in flowers in some other families (Resedaceae and Lentibulariaceae) organogenesis proceeds in exactly the opposite direction, from back to front of the apex (Tucker 1984a, b). All this is more readily understandable on Green's surface structure determining pattern (see Chapter 7), rather than some diffusion mechanism, although the latter could still perhaps be involved in determining temporal sequence.

The flower is therefore not a simple relay system, in which the formation of successive whorls of organs depends on the presence of the previous whorls, or their nature. This is implied by the observations that, for example, in the Primulaceae the petal primordia appear after the stamens (Nishino 1983) or petals and stamens are initiated as a common primordial structure (Tucker 1984b). When the initiation of petals and stamens was prevented specifically by a toxin, expressed by its gene which was spliced to *APETALA3* (a gene required for the formation of petals and stamens; Table 10.1), sepals and carpels were still formed as normal (Day, Galgoci and Irish 1995). There appears to be no requirement, therefore, for signals between successive whorls of organs. In the Leguminosae, two kinds of organs can be produced simultaneously, petal primordia being formed on one flank of the floral apex while stamens are being produced on the other (Tucker 1984b). This is also paralleled by the production of different organs simultaneously on different sides of the apex in reverting *Impatiens* (Simon 1973). It seems that the initial determination of the organs and their sites could progress

Table 10.1. *Some principal homeotic genes in* Arabidopsis *and* Antirrhinum.

	B		
	A		C
Sepals	Petals	Stamens	Carpels

	Homologous genes of		Phenotype for which wild-type gene required
Class of gene	*Arabidopsis*	*Antirrhinum*	
Affecting inflorescence	*LEAFY* *APETALA1*	*FLORICAULA* *SQUAMOSA*	flower meristem identity
	TERMINAL-FLOWER	*CENTRO-RADIALIS*	continuation of activity of inflorescence meristem
Affecting flower			
A	*APETALA2*	*OVULATA*	sepals
B	*APETALA3* *PISTILLATA*	*DEFICIENS* *GLOBOSA*	petals (with A) stamens (with C)
C	*AGAMOUS* *APETALA1*	*PLENA* *PETALOIDEA*	carpels
	SUPERMAN	–	end of formation of floral organs and end of flower meristem activity

Genes: A, sepals; A + B, petals; B + C, stamens; C, carpels. The mutant *leafy* and *floricaula* phenotypes have inflorescences but with flowers replaced by vegetative shoots. In *superman* mutants, after forming the normal number of stamens, the meristem goes on to form many more stamens, before eventually forming carpels or ceasing to function.

sequentially but that the emergence and formation of the organs follows some other pattern which does not appear to be sequential. That is, the organs are probably determined before they are visible, and the revelation of the pattern is removed in time from its original imprinting. This would imply that patterning occurs first and then is revealed later, perhaps by the realization of an unusual stress pattern leading to eventual buckling of the surface across the apex, rather than concentrically, at primordium initiation (Green 1994).

10.2. Specification of the floral organs: homeotic mutants

Screening for mutants of flower form, especially in *Arabidopsis* (Bowman 1994) and *Antirrhinum* (Bradley et al. 1993), has thrown up a number of homeotic mutations which alter the form of the flower by causing one or more floral whorls to be repeated, or missing or replaced by other organs characteristic of another whorl. Often the form of the floral organs is also modified so that organs with characteristics intermediate between whorl types are formed (Bowman, Smyth and Meyerowitz 1991). These aberrant organs remain a challenge to be fully explained by the action of the homeotic genes.

The initial transition from a leafy shoot to an inflorescence axis is controlled by the homologous genes *LEAFY* in *Arabidopsis* and *FLORICAULA* in *Antirrhinum* (Bradley et al. 1997). The phenotype of the mutants is basically an axis bearing leaves in the axils of which leafy shoots form instead of flowers, so that a 'vegetative inflorescence' is formed. These genes therefore seem to be have several effects: to eliminate internodes and to control the switching on of the genes concerned with the formation of floral rather than vegetative organs. There is a range of *LEAFY* mutants that vary in strength. With strong alleles only occasionally do secondary flowers arise in the axils of the outer organs, but when they do these flowers may show internode elongation, especially when their outermost organs are very leaf-like (Weigel et al. 1992). This is consistent with the determination of the flower coinciding with the disappearance of internodal initial cells (Lyndon 1987a). Flowers with weak or intermediate *LEAFY* alleles have more organs with petal or stamen sectors, analogous to the intermediate organs found in reverting plants (Battey and Lyndon 1990). *In situ* hybridization has shown that the *LEAFY* gene is normally expressed in the developing flower buds from the moment that they begin to form. It is not, however, expressed in the inflorescence meristem itself (Weigel et al. 1992).

The continued indeterminate growth of an inflorescence seems to be controlled by genes such as *TFL1* (*TERMINAL FLOWER*) in *Arabidopsis* and its homologue *CENTRORADIALIS* in *Antirrhinum* (Table 10.1) which are required to preserve the activity of the inflorescence meristem and to prevent it differentiating into a terminal flower. In the mutants, a terminal flower, or flowers, forms at the tip of the inflorescence, so terminating its growth (Bradley et al. 1997). *TFL1* appears to interact with *LEAFY* and *APETALA1* and *APETALA2* to regulate inflorescence development (Shannon and Meeks-Wagner 1993). It would be interesting to see if there is a *TFL1*-like gene in a plant such as *Silene*, in which the terminal meristem transforms into the first flower. In this case one might also expect a *LEAFY*-type gene to be expressed in the terminal meristem itself.

In *Arabidopsis*, the *apetala* mutants cause the formation of homeotic flowers in which petals are suppressed or modified to other organs. The *apetala1* mutant shows failure of petal initiation. Sepals become bracts with apetalous flowers in their axils and the bracts on these flowers repeat this structure, so that flowers within flowers are formed (Irish and Sussex 1990) (compare superflowering in reverted *Impatiens*; Simon 1973). The wild-type *APETALA2* gene determines the identity of the perianth organs in *Arabidopsis* (Kunst et al. 1989). The *agamous* mutant causes loss of stamens and carpels and their replacement by petals and sepals respectively, while the *apetala3* mutation causes loss of petals and stamens and their replacement by sepals and carpels respectively (Bowman, Smyth and Meyerowitz 1991). Homologous genes have been shown in *Antirrhinum* (Bradley et al. 1996).

A model (Fig. 10.1) for the action and interaction of the homeotic genes in *Arabidopsis* and *Antirrhinum* has been inferred from the effects of mutants on flower form. The genes fall into three classes, A, B and C, which, by their combinatorial action, specify the identity of the floral organs (Table 10.1) (Weigel and Meyerowitz 1994). The A genes (e.g. *APETALA2* in *Arabidopsis*, or its homologues in *Antirrhinum*) act during sepal and petal formation; B genes (e.g. *APETALA3*, *DEFICIENS*) during petal and stamen formation; and C genes (e.g. *AGAMOUS*, *PLENA*) during stamen and carpel formation. If either the A or C genes are mutant, then the remaining wild-type gene is expressed in the other's territory and time. When A alone is acting, then sepals are formed, and when C alone, carpels. When A and B are both acting then petals are formed, and when B and C, stamens. If we assume that in the *agamous* mutant the *APETALA2* gene is active throughout flower development and that similarly in the *apetala2* mutant the *AGAMOUS* gene is active throughout, then the flower form and the nature of each floral whorl is as predicted by the model (Fig. 10.1) (Meyerowitz et al. 1991). When the *agamous* homologue from *Brassica napus* was introduced transgenically into tobacco and constitutively expressed, it suppressed the A function homeotic genes (i.e. *APETALA2* homologues) and resulted in *Nicotiana* flowers with carpels, stamens, stamens, carpels, as predicted on the A, B, C model of flower formation (Mandel et al. 1992). This indicates that the

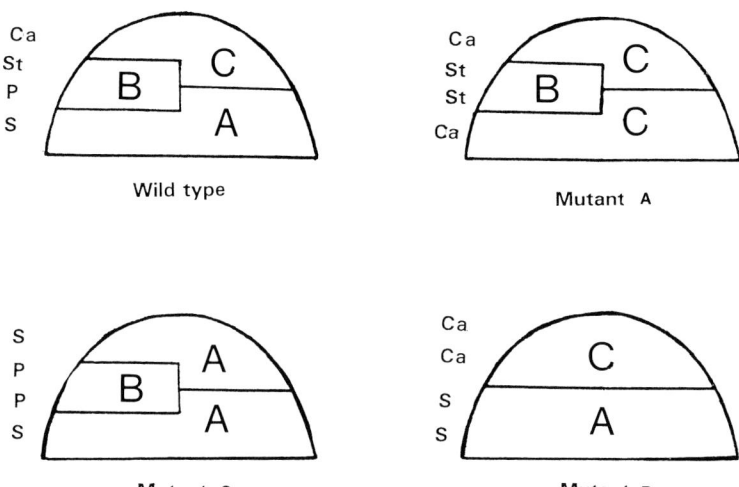

Figure 10.1 Effects of homeotic genes in the flower apex of *Arabidopsis*: the ABC model of flower development (see also Table 10.1). The normal progression of organ formation (wild-type) is sepals (S), petals (P), stamens (St) and carpels (Ca). When the A genes are acting alone, sepals are formed, and when the C genes are acting, carpels are formed. When either is not acting (i.e. mutant A or mutant C) the active gene also acts on the other's territory, producing organs in anomalous positions. The B genes act only in conjunction with the A and C genes: A + B produces petals, B + C produces stamens. When B genes are not acting (mutant B) neither petals nor stamens are formed and sepals or carpels form in their place.

homeotic genes are common to unrelated genera and function in the same way in quite different plants. Homologues of the *Arabidopsis* and *Antirrhinum* homeotic genes have been found in *Lycopersicon* (Pnueli et al. 1991) and *Petunia* (Angenent et al. 1995) and so they are assumed to be universal.

The expression of the various genes has been examined by *in situ* localization of the mRNA transcripts (Drews, Bowman and Meyerowitz 1991, Weigel et al. 1992). This has shown that in the wild-type, *AGAMOUS* RNA is present only in the stamen and carpel primordia, but that in *apetala2* mutants *AGAMOUS* is also expressed in the sepals and petals (Drews, Bowman and Meyerowitz 1991). The expression of *AGAMOUS* is therefore normally suppressed by *APETALA2*. The homeotic genes can therefore be shown as acting at specific sites and positions, and begin to be expressed when the organs that they specify are determined, just before they are initiated (Drews, Bowman and Meyerowitz 1991, Bowman et al. 1992, Goto and Meyerowitz 1994). The homeotic genes have different but overlapping spheres of activity and therefore seem to be activated according to positional information in the apical meristem. However, this may also represent a temporal sequence, because of the growth of the shoot apex as the organs are initiated in sequence, but this is questionable. The suggestion has been made that the homeotic genes are in fact misnamed and are really hetero-chronic, or timing, genes (Lord 1991). This helps to explain some features of their action that the spatial model does not. The *squamosa* mutant gene in *Antirrhinum* causes reiteration of the programme of inflorescence initiation and so 'may be regarded as a heterochronic gene' (Huijser et al. 1992, p.1247). The formation of hybrid organs, often with the base of one type, and the tip of the succeeding type (Komaki et al. 1988), and the effect of mutants such as *appendix*, which produces pistil-like outgrowths on the tips of fertile anthers (Broadhvest et al. 1992), are more readily explained as alterations in the timing of gene expression than in the position of gene expression on the meristem. However, care must be taken to distinguish between the expression of the genes at the cellular level (as RNA transcripts or as proteins) and their action at the level of determination and formation of the floral organs.

There are other genes also involved in determining flower identity. The *SUPERMAN* gene in *Arabidopsis* seems to be required to terminate action of the other homeotic genes, particularly *APETALA3*, to suppress the action of B function genes while the C function continues, and per-haps eventually to suppress C function activity. When it is in the mutant form, *superman*, the meristem continues to form stamens until it either ceases to function, or a gynoecium, often modified in structure, is formed (Bowman et al. 1992). The gene *FBP1* in *Petunia* seems to be similar (Angenent et al. 1994). Such homeotic genes that limit the action of the other genes have been christened cadastral genes (Bowman et al. 1992).

A key discovery was that the homeotic gene *AGAMOUS* in *Arabidopsis* (Yanofsky et al. 1990) and the *DEFICIENS* gene in *Antirrhinum* (Sommer et al. 1990) show DNA sequence homology with genes for DNA transcription factors isolated from yeast and man. These genes belong, therefore, to the family of regulatory genes that govern developmental switches. All the homeotic genes so far examined have a conserved region of DNA in common – the MADS box. This is so called because it is found in the gene for MCM1 protein in yeast (a transcription factor), in the genes *AGAMOUS* and *DEFICIENS* (homeotic genes in *Arabidopsis* and *Antirrhinum* repectively), and in the gene for Serum Response Factor in vertebrates (which is a DNA-binding transcriptional control factor). The inference is that the homeotic genes also code for

DNA-binding transcriptional control factors. Of course, this raises the question of what genes are the homeotic gene products controlling? Are there common control regions on all genes that are expressed in a particular floral organ?

There is evidence that at least some of the homeotic genes become expressed exclusively at primordium sites rather than being expressed equally all round the meristem annulus (Simon et al. 1994). They specify organ identity for each floral whorl, but variation in the numbers of organs per whorl, i.e. meristic variation, is probably controlled by other genes, although some homeotic genes such as *SUPERMAN* may sometimes affect organ number (Bowman, Smyth and Meyerowitz 1991). However, other genes may also be involved such as *PERIANTHA*, which also alters flower pattern (Running and Meyerowitz 1996). The number (and therefore positions) of primordia per whorl also depends on meristem size (Stevens, Huether and Wilson 1972) and so genes like *CLAVATA*, which controls organization of the meristem and causes fasciation (Leyser and Furner 1992), also affect organ number.

10.3.   Reversion of flowering

Flower development is normally determinate, the meristem ceasing activity as it is used up to form the terminal floral organs, usually carpels. However, sometimes meristematic activity can continue and the flower can revert to vegetative growth (Battey and Lyndon 1990). Such flower reversion is apparent by a leafy shoot growing out of the middle of a flower, as in *Anagallis* and *Impatiens* when the plants are transferred to non-inductive photoperiods during flower formation. Reversion in these plants can occur at any point during flower development. Since it is brought about by photoperiod, and this is perceived by the leaves, these flowers presumably require continued floral stimulus from the leaves to continue development. In flowers not able to revert, the flower shoot apex has become autonomous for development and no longer relies on signals from elsewhere in the plant.

Reversion of the inflorescence can also occur in a number of plants (Battey and Lyndon 1990). Potential flower meristems develop instead as leafy shoots or branches that do not bear flowers. Reversion is part of normal development in the pineapple, as shown by its tuft of leaves at the top of the inflorescence, and in the spider plant where, after flowering, the apex goes on to produce a small vegetative plantlet. Reversion earlier in the flowering transition leads to partial or incomplete flowering or to vegetative growth that has the branching structure of an inflorescence.

Reversion often seems to be brought about by environmental conditions that are the opposite of those inducing flowering, for instance, short-days in long-day plants and vice versa, or high temperatures in plants induced to flower by cold (Battey and Lyndon 1990). This implies that the action of the homeotic genes can be switched on or off or modified by environmental signals even after flower initiation and development has begun. This may be because such signals are mediated by plant growth substances that might in some way be able to affect the formation of dimers between the homeotic gene products and so modify or reverse the processes of floral organ determination – but this is complete speculation.

Partial flowering is similar to the effects of certain homeotic mutants and shows that some homeotic genes can be activated, without activating all. This might point to a sequence of stimuli required for full homeotic gene expression and determination of all floral organs. Alternatively, a continued (but the same) stimulus may be required for full activation of homeotic gene activity and the continued production of dimers by the early homeotic genes to keep the flowering process going, in keeping with the scheme suggested by Davies and Schwarz-Sommer (1994) (see Section 10.5).

Reversion is presumably linked to changes in the expression of genes such as *FLORICAULA, LEAFY* and *SQUAMOSA*, which are involved in the switch between vegetative and floral growth at the shoot apex. It might be expected that the expression of these genes could be controlled by changing environmental factors in the same way that reversion is (Bradley et al. 1996).

10.4.   Modification of gene action by growth substances

Homeotic mutants have been produced by screening for lesions in polyamine synthesis (Malmberg et al. 1985). Different mutants showed anthers partially converted to petals, ovules transformed into stamens (stamenoid ovules), stigmoid anthers, and the nested ('Russian doll') phenotype where flower formation starts again within the pistil of the developing flower and this structure is repeated. In the *stamenless-2*

mutant of tomato at intermediate temperatures the stamens are dis-
torted and bear external ovules (Sawhney and Greyson 1973a). At
higher temperatures (day/night: 28/23°C) the stamens do not develop
at all. At lower temperatures (18/15°C) the stamens of the mutant are
normal. Polyamine levels are higher in the mutant than in the wild-
type, and also higher at the higher temperature which enhances the
mutant phenotype (Rastogi and Sawhney 1990a). Inhibitors of polyamine synthesis can restore normal stamen development in the mutant
and application of polyamines to the wild-type can induce the formation of abnormal stamens like those in the mutant (Rastogi and
Sawhney 1990b). Auxin causes the formation of carpels instead of
stamens in the mutant, and gibberellic acid can restore the normal
phenotype to the mutants (Sawhney and Greyson 1973b), but the
action of the growth substances could be far removed from the
molecular site of gene action.

Polar auxin transport also seems to be necessary for the action of the
PINFORMED gene in *Arabidopsis*, which modifies flower form. The
pinformed mutants have wide petals, no stamens and a pistil-like structure but with no ovules in the ovary, and an auxin polar transport
activity reduced to ≤14% of normal (Okada et al. 1991). The normal
phenotype could be restored by the application of inhibitors of auxin
transport.

In the determination of sex in unisexual flowers, the action of the
homeotic genes is apparently modified by other genes, as in white campion (Hardenack et al. 1994). In the stamen-less mutants of tomato, the
initiation of the stamen primordia, and their development until they are
100μm long, was the same in the mutants as in the wild-type (Sawhney
and Greyson 1973a). Unisexuality in plants seems to be usually the
result of stamen or carpel primordia either becoming aborted soon
after initiation or simply not developing (Dellaporta and Calderon-
Urrea 1993).

Changes in flower form can be therefore be produced by (1) homeotic
genes, (2) environment, e.g. photoperiod, and (3) growth substances and
polyamines. It seems likely that various factors can affect, directly or
indirectly, the action of the homeotic genes. However, it is not known
whether plant growth substances can act directly as regulators of homeotic gene action, or what the possible transduction chains are between
environmental signals and gene action. In fact, a key question cannot
yet be answered: do homeotic genes act via plant growth substances, or
do plant growth substances act via homeotic genes?

10.5. Determination in the flower

Determination is a hierarchical process. An apex determined as floral undergoes further determination steps during flower development as the different types of floral organ are formed. The developmental state of the apex has been tested in developing flowers by bisecting them and seeing what type of floral organs formed on the cut surface. The organs that formed were always either the same as the organs being initiated on the uncut side of the apex (*Aquilegia*, Jensen 1971) or those later in the developmental sequence from those that were already present (*Primula*, Cusick 1956; *Portulaca*, Soetiarto and Ball 1969). This suggests that the apex went through a series of developmental states, or states of determination.

The sequential nature of determination in the flower meristem has also been shown by experiments with *Silene*. Plants of *Silene coeli-rosa* require 7 long-days for complete floral induction, and flowering continues if the plants are returned to non-inductive short-days (Miller and Lyndon 1976). Apices of *Silene* plants that had been given 4–7 long-days, or 7 long-days + 1 or 2 short-days, and so induced to flower to different extents, were excised and cultured on a basal medium. Apices that had become determined for the formation of a particular floral whorl developed this whorl in the same way as intact plants receiving the same inductive treatment. If the apices had been excised before determination was complete, then they did not develop this whorl as the apices on the intact plants did. In this way it was shown that the presumed time taken for determination was 2 days for the sepals, 3 days for the petals and stamens 1–5, and 4 days for stamens 6–10 and the carpels (Donnison and Francis 1993).

Excised, cultured, apices could also be made to revert to vegetative growth after the initiation of each of the floral whorls had been formed (Donnison and Francis 1994). If a particular type of floral organ formed, then the apex was determined for that particular organ type. By excising the apices at daily intervals after floral induction and seeing whether they reverted, it was confirmed that determination for each successive floral organ type took about 3 days. This means that each of the homeotic genes are determined separately and sequentially rather than there being one determination event for the whole set of flowering genes.

Until recently determination was only a physiological concept, of which the cellular basis remained a mystery. Exciting work has revealed a mechanism at the molecular level which seems to show how gene expression becomes locked into a new mode and could also explain why determination is gradual in nature. In *Antirrhinum* the homeotic genes *DEFICIENS* and *GLOBOSA* produce proteins that can form heterodimers which act as feedback promoters of the transcription of the genes (Fig. 10.2), so that transcription of both genes becomes progressively reinforced (Schwarz-Sommer et al. 1992, Tröbner et al. 1992, Davies and Schwarz-Sommer 1994). These interactions have been confirmed by *in situ* analyses of mRNA and protein expression patterns (Zachgo et al. 1995). This locking-in of gene transcription is in fact determination at the molecular level and could be essentially what determination is all about. The action of dimers as promoters is specific. In *Arabidopsis*, the binding ability of different dimers to a DNA sequence (the CArG box) was tested. This CArG box is a sequence to which other MADS box gene products, known to be transcription factors, will bind. The only combinations capable of binding to the CArG box sequences were APETALA1 homodimers, AGAMOUS homodimers and APETALA3/PISTILLATA heterodimers (Riechmann, Krizek and Meyerowitz 1996). These are precisely the combinations predicted by the model for the action of the homeotic genes in determining floral organ identity (Fig. 10.1).

10.6. Cellular differences between floral organs

The MADS box genes, which determine floral organ identity, code for protein transcription factors which must regulate the downstream genes that program the form and function of the floral organs. Some of these downstream genes will be different in different organs, so detectable qualitative differences in gene products, i.e. ultimately the protein complement, between different organs might be expected.

10.6.1. Protein complement

Differences in protein composition of different floral organs were first shown by PAGE in the tulip (Barber and Steward 1968). The floral organs had different protein profiles, the clearest differences being between the stamens and ovaries. Floral organs also differed in their isozyme patterns for malate dehydrogenase and, to a lesser extent, for esterase. In tomato there were greater differences in protein complements between the parts of the pistil (stigma, style and ovary) than there were

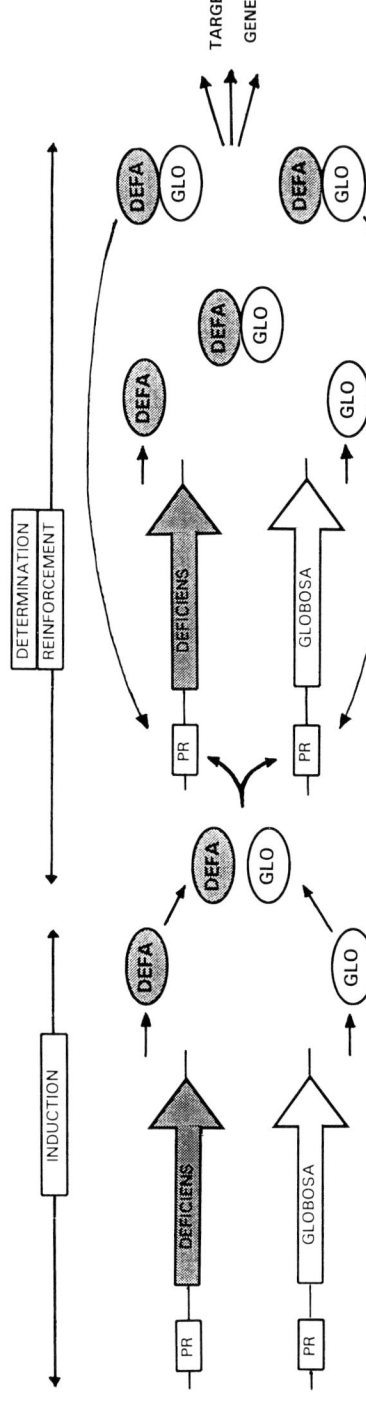

Figure 10.2 A model for determination at the molecular level. The induction of the *DEFICIENS* and *GLOBOSA* genes of *Antirrhinum* (large arrows) produces small amounts of the proteins DEFA and GLO, which combine to form heterodimers. In the reinforcement, or determination, stage these heterodimers bind to upstream promoter regions (**PR**) thus promoting transcription of the genes, which in turn promotes protein and heterodimer production. This could be the process that commits the cell to the developmental pathway, i.e. causes determination, as a result of which the downstream target genes in the developmental pathway become activated. From Davies and Schwarz-Sommer (1994).

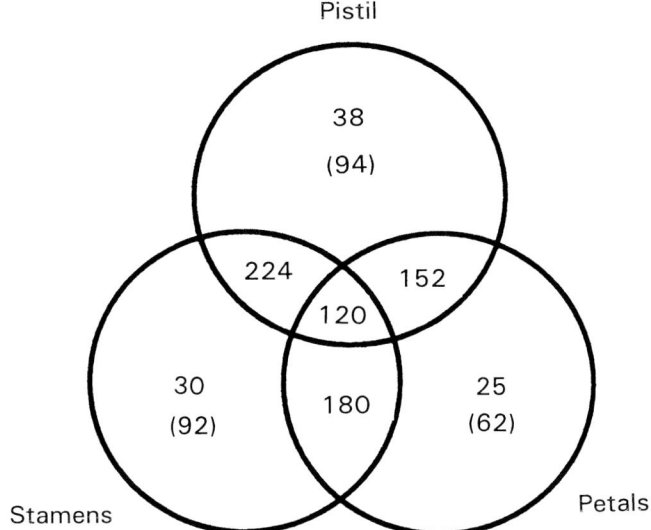

Figure 10.3 Numbers of polypeptides separated by two-dimensional polyacrylamide gel electrophoresis from extracts of floral organs of *Pharbitis* (Japanese Morning Glory). In parentheses: polypeptides unique to that floral organ but also found in stems, leaves or sepals. From Bassett, Mothershed and Galau (1988).

between ovaries and stamens (Sawhney, Chen and Sussex 1985). In the floral organs of *Pharbitis*, the number of polypeptide spots, detected by 2-D PAGE, varied from 209 for the petal extracts to 479 for the pistil (Bassett, Mothershed and Galau 1988). Of the total of about 800 identifiable spots, only about 4%, or about 30 in each floral organ (petal, stamen, pistil), were unique to each organ (Fig. 10.3). Sepals shared a greater proportion of spots with leaves than did petals, stamens or pistils, and these latter three seemed to form a distinct subset. Presumably the relatively low proportion of organ-specific polypeptides explains why they were not found in the one-dimensional gels of Solanaceae flower parts (Bhadula and Sawhney 1989). In the maize ear, again > 800 polypeptide spots were found on 2D-PAGE electropherograms, but only 39 showed quantitative changes during early organogenesis (Thompson-Coffe et al. 1992).

Many fewer proteins are detected immunologically. In *Sinapis alba* only three antigens specific to the flowers were detected (Jacqmard, Lyndon and Salmon 1984; see Chapter 9). Using protein extracts from whole tobacco flowers, 11 antibodies were raised that could be shown to react to floral proteins, but only a few of these were totally organ-specific

(Evans et al. 1988). Those with the greatest differential reactivity were those targetted against carbohydrate moieties, suggesting a possible general role for glycosylation in plant differentiation.

The total concentration of protein in the sepals and petals of solanaceous flowers was about half or less of that in the stamens and ovaries, which were similar to the young leaves. This probably reflects the highly vacuolate nature of most of the cells in the mature sepals and petals, in contrast with the highly cytoplasmic cells in the reproductive organs (Bhadula and Sawhney 1989).

10.6.2.   Gene expression

The differences in protein composition of different floral organs (see above) are evidence for changes in gene expression but do not reveal at what level between the gene and the protein that control is being exerted. Direct evidence for changes in gene expression at or near the level of transcription come from experiments with cDNA. Tissue-specific DNA has been isolated from tomato pistils (Gasser et al. 1989). Nine cDNA clones were characterized from immature (premeiotic to early meiosis) and mature (at anthesis) pistils. Four of these were also expressed in anthers. *In situ* hybridization (Smith, Hinchee and Horsch 1987) showed that the expression of these genes was regulated very tightly both in time and space. One gene was expressed only in the transmitting tissue of the styles. A second gene was expressed only in two to three cell layers of the ovules for < 8 days (Gasser et al. 1989). In the orchid *Phalaenopsis*, seven cDNAs were identified that were either stage- or tissue-specific during ovule development (Nadeau et al. 1996). The gene *AINTEGUMENTA* is required in *Arabidopsis* for ovule and female gametophyte development, and is related to *APETALA2* (Klucher et al. 1996). Although its effects are only on late development, it begins to be expressed in the early growth of all shoot and floral primordia (Elliott et al. 1996).

Similarly, genes specific to stamens have also been identified. Five anther-specific cDNA transcripts were detected in developing tomato anthers. Transcripts increased during gametogenesis and reached maximal levels in the mature pollen grains. Since transcripts also persisted in growing pollen tubes towards their growing tips and were also in the anther wall, this showed that the genes were expressed in both gametophyte- and sporophyte-derived tissues (Ursin, Yamaguchi and McCormick 1989). A clone, FA3, which increased in concentration during the transition to flowering in tobacco apical meristems (Kelly et al. 1990), was in parenchyma cells of the anther and carpel but was not in the developing pollen sac or the placenta and so seems to be concerned with the development of the reproductive organs but not with gametogenesis.

Gene expression is also tissue- and time-specific in stamens. In *Sinapis* a group of seven cDNA transcripts, confined to the tapetum of the anthers, appeared 10 days after the beginning of floral induction and declined after 15 days (Melzer, Majewski and Apel 1990). Other anther-specific cDNA clones, in sunflower, are expressed only in the anther epidermis (Evrard et al. 1991). Indeed, not unsurprisingly, there appear to be several sets of anther-specific genes each expressed in a different tissue of the anther and at different times during anther development. One of the genes involved, *TA29*, has been shown to be regulated primarily at the transcriptional level and its expression regulated by a 122 base pair $5'$ region upstream of the gene DNA (Koltunow et al. 1990).

Gene expression in developing petals is starting to be examined, especially with respect to pigment synthesis (Martin et al. 1991). Four genes involved in the biosynthesis of anthocyanin in *Antirrhinum* flowers are expressed only in the epidermis (as, too, is the anthocyanin) but all in the same pattern across the flower, although they may be expressed at slightly different times (Jackson, Roberts and Martin 1992). Although these genes are concerned with the same biosynthetic pathway and appear to be expressed coordinately, there are presumably fine controls of gene expression which remain to be discovered. There must be many other enzymes and proteins specific for petals, as there must also be for other floral organs, and especially the stamens and carpels. The molecular differentiation of floral organs has begun to be investigated and is summarized by Gasser (1991). The pattern and control, temporally and spatially, of gene activation during flower development remains a major area for future research.

The domains of gene expression do not necessarily coincide with the patterns of the effects of gene action, the functional domains. The *lateral suppressor* mutation in tomato suppresses petal formation (without altering the positions of the other floral organs) (Szymkowiak and Sussex 1993). However, the gene expression has to be in apical layers L2 or L3, since chimeras which show the mutation only in L1 have the wild-type phenotype. This parallels the observation that it is the L3 layer that determines the size of the floral meristem and the number of carpels in tomato periclinal chimeras (Szymkowiak and Sussex 1992). The homeotic gene *FLORICAULA* in *Antirrhinum* when expressed in only a single cell layer in chimeras activates downstream homeotic genes in all layers (Carpenter and Coen 1995, Hantke, Carpenter and Coen 1995). The implication is that such non-cell autonomous gene action is mediated by the movement of signal molecules from the cells where the gene is transcribed to the cells where its effects are seen. In the case of *DEFICIENS* and *GLOBOSA* (B class genes in *Antirrhinum*, Table 10.1) it is the protein products which move from cell to cell to result in non-cell autonomous gene function. Moreover, the protein movement is polar since *DEFICIENS* acts cell-autonomously when expressed only in the L1 layer and allows petal expansion to proceed, but not when it is expressed only in L2 and L3 when the epidermal cells become petalloid but the petals do not expand (Perbal et al. 1996). This incidentally is another pointer to the importance of the epidermis in determining organ form.

10.6.3. Cell division and cell cycle in floral organ primordia

Culture experiments show that young floral organs can be excised and grown to maturity without the addition of growth substances and so they seem to be potentially independent of the rest of the plant from early development (Hicks 1980). Only in plants showing flower reversion does primordium development seem to depend on stimuli from elsewhere in the plant (Battey and Lyndon 1990). Autonomy of the different floral whorls is also shown with respect to their growth and division rates. In *Silene,* the cell cycle apparently fluctuates during the initiation of the floral whorls (see Table 7.1), being faster when sepals and stamens are formed and slower when petals and carpels are forming (Lyndon and Cunninghame 1986). Within each floral whorl the same value for the cell cycle was obtained whether at primordium initiation or during the early growth of the primordia (Lyndon 1979a). This implies that in the *Silene* flower the cell cycle is fixed for a particular organ type at initiation and continues unchanged as the organ develops. This also implies that there are differences in the length of the cell cycle over distances of only a few cells on the apical surface where the floral primordia are initiated.

How do such precise differences arise in the first place? The increased rate of cell division that occurs in *Silene* apices just before sepal initiation, is indicated by mitoses principally on the flanks of the apex where the sepals will form rather than at the summit of the apex (Nougarède, Francis and Rondet 1991), but the divisions seem to be all round the apex and not clustered at the positions of the individual sepals (Francis 1992).

Are differences in division rates in some way the result of the formation of different organ types? The formation of floral organs seems to require cell division components or functions, coded by the *TSO1* gene, that are not required elsewhere in the plant (Liu, Running and Meyerowitz 1997), and it is possible that different types of floral organ also have requirements for different cell division functions. What happens to the cell cycle in an organ which starts out as one thing and finishes up as another, e.g. the petal to leaf transition in reverted, but reflowering, *Impatiens* (Battey and Lyndon 1986), or organs which are part leaf, part petal (Battey and Lyndon 1988)? The relationships between the cell cycle and development in the flower are no clearer than they are in most other developmental situations in the plant.

10.7. Flower development

The nature of the signal that commits the apex to flowering now appears to be much more complex than a single florigenic substance. But what is the role of a floral stimulus in altering patterning? Does it activate *FLORICAULA* and *LEAFY*, and homologous genes in other plants? Does it also act to cause expression of all the flower development switching genes, or does the floral stimulus just initiate a flowering cascade? In *Impatiens* a continuous supply of floral stimulus from the leaves seems to be required for flower development (Battey and Lyndon 1984, 1990) but this implies that continued floral stimulus from the leaves is not required for plants not showing reversion and that evocation is the transfer of control to the apex itself. This raises the question of the nature of the stimulus required for full flower development. Does the floral stimulus cause the production autonomously of some stimulus in the apex which is needed for floral organ development?

Some of these problems are now amenable to attack from the genetic and molecular level, resulting from the discovery of the homeotic, and related, genes, and this is now a very active area of shoot apex research. The floral stimulus activates the homeotic genes, which in turn regulate the downstream activation of further genes controlling primordium, organ and meristem development. The upstream and downstream mechanisms on either side of these switches remain as black boxes. In particular, the mechanisms and growth processes, by which form is produced, are far from being fully understood. It seems likely that genes and gene products interact in a fashion that may not be obvious from knowing the function of every gene, but that the system is one that is emergent from the interactions of its components in ways that will not be easy to predict without sophisticated models of the sorts proposed by Meinhardt (1982) and others (Goodwin and Saunders 1989) at the biochemical level, and by Green (1996) at the biophysical level.

The cellular basis for the development of form, and its manifestation at the tissue level, is clearly a lot more complex than can be explained by the action of a few homeotic genes. These are the switches. The tasks ahead are to investigate what activates the switches and to understand the nature of the machine that is switched on. The shoot apex and the primordia it produces are a unique system in which to study the challenging problems of plant morphogenesis.

References

Abbe, E.C. and Phinney, B.O. 1951. The growth of the shoot apex in maize: external features. *American Journal of Botany* 38:737–744.

Abbe, E.C., Phinney, B.O. and Baer, D.F. 1951. The growth of the shoot apex in maize: internal features. *American Journal of Botany* 38:744–751.

Abbe, E.C., Randolph, L.F. and Einset, J. 1941. The developmental relationship between shoot apex and growth pattern of leaf blade in diploid maize. *American Journal of Botany* 28:778–784.

Alberts, B., Bray, D., Lewis, J., Raff, M., Roberts, K. and Watson, J.D. 1994. *Molecular Biology of the Cell* (3rd edition). New York and London: Garland.

Angenent, G.C., Busscher, M., Franken, J., Dons, H.J.M. and van Tunen, A.J. 1995. Functional interaction between the homeotic genes *fbp1* and *pMADS1* during *Petunia* organogenesis. *Plant Cell* 7:507–516.

Angenent, G.C., Franken, J., Busscher, M., Weiss, D. and van Tunen, A.J. 1994. Co-suppression of the Petunia homeotic gene *fbp2* affects the identity of the generative meristem. *Plant Journal* 5:33–44.

Araki, T. and Komeda, Y. 1990. Electrophoretic analysis of florally-evoked meristems of *Pharbitis nil* Choisy cv. Violet. *Plant and Cell Physiology* 31:137–144.

Auderset, G., Gahan, P.B., Dawson, A.L. and Greppin, H. 1980. Glucose-6-phosphate dehydrogenase as an early marker of floral induction in shoot apices of *Spinacia oleracea* var. Nobel. *Plant Science Letters* 20:109–113.

233

Auderset, G., Gahan, P., Oniya, G.O.C. and Greppin, H. 1985. Increased pentose phosphate pathway activity linked to floral induction in apices of *Spinacia oleracea* during short days. *Annals of Botany* 55:61–64.

Ball, E. 1944. The effects of synthetic growth substances on the shoot apex of *Tropaeolum majus* L. *American Journal of Botany* 31:316–327.

Ball, E. 1952a. Experimental division of the shoot apex of *Lupinus albus* L. *Growth* 16:151–174.

Ball, E. 1952b. Morphogenesis of shoots after isolation of the shoot apex of *Lupinus albus*. *American Journal of Botany* 39:167–191.

Ball, E. 1960a. Cell division in living shoot apices. *Phytomorphology* 10:377–396.

Ball, E. 1960b. Sterile culture of the shoot apex of *Lupinus albus*. *Growth* 24:91–110.

Ball, E. and Boell, E. J. 1944. Respiratory rates of the shoot tips and maturing tissues in *Lupinus albus* and *Tropaeolum majus*. *Proceedings of the National Academy of Sciences, USA* 30:45–50.

Ball, E. and Soma, K. 1965. Effect of sugar concentration on growth of the shoot apex of *Vicia faba*. In *Proceedings of an International Conference on Plant Tissue Culture*, eds P.R. White and A.R. Grove, pp. 269–285. Berkeley: McCutchan.

Ballade, P. 1970. Precisions nouvelles sur la caulogenese apicale des racines axil-laites du cresson (*Nasturtium officinale* R. Br.). *Planta* 92:138–145.

Barber, J.T. and Steward, F.C. 1968. The proteins of *Tulipa* and their relation to morphogenesis. *Developmental Biology* 17:326–349.

Barton, M.K. and Poethig, R.S. 1993. Formation of the shoot apical meristem in *Arabidopsis thaliana*: an analysis of development in the wild type and in the *shoot meristemless* mutant. *Development* 119:823–831.

Basile, D.V. 1967. The influence of hydroxy-L-proline on ontogeny and morpho-genesis of the liverwort, *Scapania nemorosa*. *American Journal of Botany* 54:977–983.

Basile, D.V. 1990. Morphoregulatory role of hydroxyproline-containing proteins in liverworts. In *Bryophyte Development: Physiology and Biochemistry*, ed. R.N. Chopra, pp. 225–243. Boca Raton: CRC Press.

Basile, D.V. and Basile, M.R. 1983. Desuppression of leaf primordia of *Plagiochila arctica* (Hepaticae) by ethylene antagonist. *Science* 220:1051–1053.

Bassett, C.L., Mothershed, C.P. and Galau, G.A. 1988. Floral-specific polypep-tides in the Japanese morning glory. *Planta* 175:221–228.

Battey, N.H. and Lyndon, R.F. 1984. Changes in the apical growth and phyllo-taxis on flowering and reversion in *Impatiens balsamina* L. *Annals of Botany* 54:553–567.

Battey, N.H. and Lyndon, R.F. 1986. Apical growth and modification of the development of primordia during re-flowering of reverted plants of *Impatiens balsamina* L. *Annals of Botany* 58:333–341.

Battey, N.H. and Lyndon, R.F. 1988. Determination and differentiation of leaf and petal primordia in *Impatiens balsamina* L. *Annals of Botany* 61:9–16.

Battey, N.H. and Lyndon, R.F. 1990. Reversion of flowering. *Botanical Review* 56:162–189.

Bell, M.H., Halford, N.G., Ormrod, J.C. and Francis, D. 1993. Tobacco plants transformed with *cdc25*, a mitotic inducer gene from fission yeast. *Plant Molecular Biology* 23:445–451.

Bennett, M.D. 1976. The cell in sporogenesis and spore development. In *Cell Division in Higher Plants*, ed. M.M. Yeoman, pp. 161–198. London: Academic Press.

Berg, A.R. and Cutter, E.G. 1969. Leaf initiation rates and volume growth rates in the shoot apex of *Chrysanthemum*. *American Journal of Botany* 56:153–159.

Bergann, F. 1965. Wächst *Epilobium* mit Scheitelzellen? *Berichte der Deutschen Botanischen Gesellschaft* 78:405–410.

Bernier, G. 1964. Etude histophysiologique et histochimique de l'évolution du méristème apical de *Sinapis alba* L., cultivé en milieu conditionné et diverses durées de jour favorables ou défavorables à la mise a fleurs. Doctoral thesis, University of Liège.

Bernier, G. 1969. *Sinapis alba* L. In *The Induction of Flowering: Case Histories*, ed. L.T. Evans, pp. 305–327. Sydney: Macmillan.

Bernier, G. 1971. Structural and metabolic changes in the shoot apex in transition to flowering. *Canadian Journal of Botany* 49:803–819.

Bernier, G. 1979. The sequences of floral evocation. In *La Physiologie de la Floraison*, eds P. Champagnat and R. Jacques, pp. 129–168. Paris: CNRS.

Bernier, G. 1984. The factors controlling evocation: an overview. In *Light and the Flowering Process*, eds D. Vince-Prue, B. Thomas and K.E. Cockshull, pp. 277–292. London: Academic Press.

Bernier, G. 1988. The control of floral evocation and morphogenesis. *Annual Review of Plant Physiology* 39:175–219.

Bernier, G. and Bronchart, R. 1963. Application de la technique d'histoautoradiographie à l'étude de l'incorporation de thymidine tritiée dans les méristèmes caulinaires. *Bulletin de la Société Royale de Sciences Liège* 32:269–283.

Bernier, G., Bronchart, R. and Jacqmard, A. 1964. Action of gibberellic acid on the mitotic activity of different zones of the shoot apex of *Rudbeckia bicolor* and *Perilla nankinensis*. *Planta* 61:236–283.

Bernier, G., Havelange, A., Houssa, C., Petitjean, A. and Lejeune, P. 1993. Physiological signals that induce flowering. *Plant Cell* 5:1147–1155.

Bernier, G., Kinet, J.-M. and Bronchart, R. 1967. Cellular events at the meristem during floral induction in *Sinapis alba* L. *Physiologie Végétale* 5:311–324.

Bernier, G., Kinet, J.-M., Bodson, M., Rouma, Y. and Jacqmard, A. 1974. Experimental studies on the mitotic activity of the shoot apical meristem and its relation to floral evocation and morphogenesis in *Sinapis alba*. *Botanical Gazette* 135:345–352.

Bernier, G., Kinet, J.-M., Jacqmard, A., Havelange, A. and Bodson, M. 1977. Cytokinin as a possible component of the floral stimulus in *Sinapis alba*. *Plant Physiology* 60:282–285.

Bernier, G., Kinet, J.-M. and Sachs, R.M. 1981. *The Physiology of Flowering*, vol. II. Boca Raton: CRC Press.

Berthier, J. 1972. Recherches sur la structure et le développement de l'apex du gametophyte feuillé des Mousses. *Révue Bryologique et Lichenologique* 38:421–551.

Besnard-Wibaut, C., Noin, M. and Zeevaart, J.A.D. 1983. Mitotic activities and levels of nuclear DNA in the apical meristem of *Silene armeria* (strain S1.2) following application of gibberellin A3. *Plant and Cell Physiology* 24:1269–1279.

Bhadula, S.K. and Sawhney, V.K. 1989. Protein analysis of floral organs of some members of Solanaceae. *Botanical Magazine Tokyo* 102:85–91.

Bhatla, S.C. and Chopra, R.N. 1984. Subcellular localization of adenylate cyclase in the shoot apices of *Bryum argentum* Hedw. *Annals of Botany* 54:195–200.

Bierhorst, D.W. 1959. Symmetry in *Equisetum*. *American Journal of Botany* 46:170–179.

Bierhorst, D.W. 1977. On the stem apex, leaf initiation and early leaf ontogeny in filicalean ferns. *American Journal of Botany* 64:125–152.

Bodson, M. 1970. Influence d'un abaissement de la température et d'un retour à température ordinaire sur l'activité mitotique du méristème caulinaire de *Sinapis alba*. *C.R. Acad. Sci. Paris D* 270:314–316.

Bodson, M. 1975. Variation in the rate of cell division in the apical meristem of *Sinapis alba* during transition to flowering. *Annals of Botany* 39:547–554.

Bodson, M. 1984. Assimilates and evocation. In *Light and the Flowering Process*, eds D. Vince-Prue, B. Thomas and K.E. Cockshull, pp. 157–169. London: Academic Press.

Bodson, M. and Outlaw, W.H. 1985. Elevation in the sucrose content of the shoot apical meristem of *Sinapis alba* at floral evocation. *Plant Physiology* 79: 420–424.

Boke, N.H. 1940. Histogenesis and morphology of the phyllode in certain species of *Acacia*. *American Journal of Botany* 27:73–90.

Bonnot, E.-J. 1967. Sur la structure de l'apex du gamétophyte feuillé de la mousse *Anomodon viticulosus* (L.). *Bulletin de la Société Botanique de France* 114:4–11.

Bowes, B.G. 1965. The ultrastructure of the shoot apex and young shoot of *Glechoma hederacea* L. *La Cellule* 65:351–356.

Bowman, J. (ed.) 1994. *Arabidopsis: An Atlas of Morphology and Development*. Springer, New York.

Bowman, J.L., Drews, G.N. and Meyerowitz, E.M. 1991. Expression of the *Arabidopsis* floral homeotic gene AGAMOUS is restricted to specific cell types late in flower development. *Plant Cell* 3:749–758.

Bowman, J.L., Sakai, H., Jack, T., Weigel, D., Mayer, U. and Meyerowitz, E.M. 1992. *SUPERMAN*, a regulator of floral homeotic genes in *Arabidopsis*. *Development* 114:599–615.

Bowman, J.L., Smyth, D.R. and Meyerowitz, E.M. 1991. Genetic interactions among floral homeotic genes of *Arabidopsis*. *Development* 112:1–20.

Bradley, D., Carpenter, R., Elliott, R., Simon, R., Romero, J., Hantke, S., Doyle, S., Mooney, M., Luo, D., McSteen, P., Copsey, L., Robinson, C. and Coen, E. 1993. Gene regulation of flowering. *Philosophical Transactions of the Royal Society of London B* 339:193–197.

Bradley, D., Ratcliffe, O., Vincent, C., Carpenter, R. and Coen, E. 1997. Inflorescence commitment and architecture in *Arabidopsis*. *Science* 275:80–83.

Bradley, D., Vincent, C., Carpenter, R. and Coen, E. 1996. Pathways for inflorescence and floral induction in *Antirrhinum*. *Development* 122:1535–1544.

Broadhvest, J., Daigle, N., Martin, M., Haughn, G.W. and Bernier, F. 1992. Appendix: a novel type of homeotic mutation affecting floral morphology. *Plant Journal* 2:991–997.

Bronchart, R. and Nougarède, A. 1970. Evènements métaboliques et structuraux au niveau du méristème du 'Perilla nankensis' lors de la phase préflorale. In *Cellular and Molecular Aspects of Floral Induction*, ed. G. Bernier, pp. 27–36. London, Longman.

Brown, R. 1951. The effects of temperature on the durations of the different stages of cell division in the root tip. *Journal of Experimental Botany* 2:96–110.

Burn, J.E., Bagnall, D.J., Metzger, J.D., Dennis, E.S. and Peacock, W.J. 1993. DNA methylation, vernalization and the initiation of flowering. *Proceedings of the National Academy of Sciences, USA* 90:287–291.

Butler, R.D. and Lane, G.R. 1959. The study of apical development in relation to etiolation. *Journal of the Linnaean Society, London* 56:170–176.

Cannell, M.G.R. 1976. Shoot apical growth and cataphyll initiation rates in provenances of *Pinus contorta* in Scotland. *Canadian Journal of Forest Research* 6:539–556.

Cannell, M.G.R. 1978. Analysis of shoot apical growth of *Picea sitchensis* seedlings. *Annals of Botany* 42:1291–1303.

Cannell, M.G.R. and Cahalan, C.M. 1979. Shoot apical meristems of *Picea sitchensis* seedlings accelerate in growth following bud-set. *Annals of Botany* 44:209–214.

Cannell, M.G.R. and Willett, S.C. 1975. Rates and times at which needles are initiated in buds on differing provenances of *Pinus contorta* and *Picea sitchensis* in Scotland. *Canadian Journal of Forest Research* 5:367–380.

Carpenter, R. and Coen, E.S. 1995. Transposon induced chimeras show that *floricaula*, a meristem identity gene, acts non-autonomously between cell layers. *Development* 121:19–26.

Carpenter, R., Copsey, L., Vincent, C., Doyle, S., Magrath, R. and Coen, E. 1995. Control of flower development and phyllotaxy by meristem identity genes in *Antirrhinum*. *Plant Cell* 7:2001–2011.

Caruso, J.L. 1968. Morphogenetic aspects of a leafless mutant in tomato. I. General patterns in development. *American Journal of Botany* 55:1169–1176.

Cavalier-Smith, T. 1985. Genetic and epigenetic control of the plant cell cycle. In *The Cell Division Cycle in Plants*, (SEB Seminar Series 26), eds J.A. Bryant and D. Francis, pp.179–197. Cambridge, Cambridge University Press.

Cecich, R.A. 1977. An electron microscopic evaluation of cytohistological zonation in the shoot apical meristem of *Pinus banksiana*. *American Journal of Botany* 64:1263–1271.

Cecich, R.A., Lersten., N.R. and Miksche, J.P. 1972. A cytophotometric study of nucleic acids and proteins in the shoot apex of white spruce. *American Journal of Botany* 59:442–449.

Chai, H. 1990. Buckling and post-buckling behaviour of elliptical plates. Part II – Results. *Journal of Applied Mechanics* 57:989–994.

Chailakhyan, M.Kh., Aksenova, N.P., Konstantinova, T.N. and Bavrina, T.V. 1974. Use of tobacco stem calluses for the investigation of some regularities of flowering. *Phytomorphology* 24:86–96.

Chapman, J.M. and Perry, P. 1987. A diffusion model of phyllotaxis. *Annals of Botany* 60:377–389.

Charles-Edwards, D.A., Cockshull, K.E., Horridge, J.S. and Thornley, J.H.M. 1979. A model of flowering in *Chrysanthemum*. *Annals of Botany* 44:557–566.

Charlton, W.A. 1977. Evaluation of sequence and rate of lateral-root initiation in *Pontederia cordata* L. by means of colchicine inhibition of cell division. *Botanical Gazette* 138:71–79.

Christianson, M.L. 1996. Morphogenesis and the coordination of cell division in the bryophytes. *Seminars in Cell and Developmental Biology* 7:881–889.

Clark, A.M., Verbeke, J.A. and Bohnert, H.J. 1992. Epidermis-specific gene expression in *Pachyphytum*. *Plant Cell* 4:1189–1198.

Clark, S.E., Running, M.P. and Meyerowitz, E.M. 1993. CLAVATA1, a regulator of meristem and flower development in *Arabidopsis*. *Development* 119:397–418.

Clowes, F.A.L. 1961. *Apical Meristems*. Oxford: Blackwell.

Clowes, F.A.L. 1976. The root apex. In *Cell Division in Higher Plants*, ed. M.M. Yeoman, pp. 253–284. London: Academic Press.

Clowes, F.A.L. and MacDonald, M.M. 1987. Cell cycling and the fate of potato buds. *Annals of Botany* 59:141–148.

Coen, E.S. 1991. The role of homeotic genes in flower development and evolution. *Annual Review of Plant Physiology and Plant Molecular Biology* 42:241–279.

Corson, G.E. 1969. Cell division studies of the shoot apex of *Datura stramonium* during transition to flowering. *American Journal of Botany* 56:1127–1134.

Cottignies, A. 1974. Les critères nucléaires d'une dormance vraie et totale dans le point végétatif du *Fraxinus excelsior* L. *Planta* 120:171–179.

Cottignies, A. 1977. Le nucléole dans le point végétatif dormant et non dormant du *Fraxinus excelsior* L. *Zeitschrift für Pflanzenphysiologie* 83:189–200.

Cottignies, A. 1979. The blockage in the G1 phase of the cell cycle in the dormant shoot apex of ash. *Planta* 147:15–19.

Cottignies, A. 1983. Dormance totale et distribution normale des teneurs en ADN dans le point végétatif du Frêne. *C.R. Acad. Sci. Paris D* 296:829–832.

Cottrell, J.E. and Dale, J.E. 1986. The effects of photoperiod and treatment with gibberellic acid on the concentration of soluble carbohydrates in the shoot apex of spring barley. *New Phytologist* 102:365–373.

Cremer, F. 1992. Etude des changements dans l'expression du génome pendant l'induction photopériodique et la transition florale chez *Sinapis alba*. Doctoral thesis, University of Liège.

Cremer, F., van de Walle, C. and Bernier, G. 1985. Two-dimensional gel electrophoresis of polypeptides from vegetative and reproductive buds of *Sinapis alba*. *Archives Internationales de Physiologie et de Biochimie* 94:9–10.

Cremer, F., van de Walle, C. and Bernier, G. 1992. Changes in the pattern of proteins synthesized in the shoot apical meristem of *Sinapis alba* during floral transition. *Plant and Cell Physiology* 33:1199–1207.

Crockett, L.J. 1968. The effects of chronic gamma irradiation on the internal apical configurations of the vegetative shoot apex of *Coleus blumei*. *American Journal of Botany* 55:265–268.

Croxdale, J.G. 1983. Quantitative measurements of phosphofructokinase in the shoot apical meristem, leaf primordia and leaf tissues of *Dianthus chinensis* L. *Plant Physiology* 73:66–70.

Croxdale, J.G. and Outlaw, W.H. 1983. Glucose-6-phosphate-dehydrogenase activity in the shoot apical meristem, leaf primordia and leaf tissues of *Dianthus chinensis* L. *Planta* 157:289–297.

Croxdale, J.G. and Vanderveer, P.J. 1986. Quantitative measurements of hexo-kinase activity in the shoot apical meristem, leaf primordia and leaf tissues of *Dianthus chinensis* L. *Plant Physiology* 81:186–191.

Cunninghame, M.E. and Lyndon, R.F. 1986. The relationship between the dis-tribution of periclinal cell divisions in the shoot apex and leaf initiation. *Annals of Botany* 57:737–746.

Cusick, F. 1956. Studies of floral morphogenesis. I. Median bisections of flower primordia in *Primula bulleyana* Forrest. *Transactions of the Royal Society of Edinburgh* 63:153–166.

Cutter, E.G. 1954. Experimental induction of buds from fern leaf primordia. *Nature* 173:440–441.

Cutter, E.G. 1959. On a theory of phyllotaxis and histogenesis. *Biological Reviews* 34:243–263.

Cutter, E.G. 1965. Recent experimental studies of the shoot apex and shoot morphogenesis. *Botanical Review* 31:7–113.

Cutter, E.G. 1971. *Plant Anatomy: Experiment and Interpretation*. Part 2 Organs. London: Arnold.

D'Amato, F. and Avanzi, S. 1968. The shoot apical cell of *Equisetum arvense*, a quiescent cell. *Caryologia* 21:83–89.

Davies, B. and Schwarz-Sommer, Z. 1994. Control of floral organ identity by homeotic MADS-box transcription factors. In *Plant Promoter and Transcription Factors*, ed. L. Nover, pp. 235–258. Berlin: Springer.

Davis, E.L., Rennie, P. and Steeves, T.A. 1979. Further analytical and experi-mental studies on the shoot apex of *Helianthus annuus*: variable activity in the central zone. *Canadian Journal of Botany* 57:971–980.

Day, C.D., Galgoci, B.F.C. and Irish, V.F. 1995. Genetic ablation of petal and stamen primordia to elucidate cell interactions during floral development. *Development* 121:2887–2895.

Dellaporta, S.L. and Calderon-Urrea, A. 1993. Sex determination in flowering plants. *Plant Cell* 5:1241–1251.

Denne, M.P. 1966a. Morphological changes in the shoot apex of *Trifolium repens* L. 1. Changes in the vegetative apex during the plastochron. *New Zealand Journal of Botany* 4:300–314.

Denne, M.P. 1966b. Diurnal and plastochronal changes in the shoot apex of *Tradescantia fluminensis* Vell. *New Zealand Journal of Botany* 4:444–454.

Denne, M.P. 1966c. Morphological changes in the shoot apex of *Trifolium repens* L. 2. Diurnal changes in the vegetative apex. *New Zealand Journal of Botany* 4:434–443.

Dennin, K.A. and McDaniel, C.N. 1985. Floral determination in axillary buds of *Nicotiana sylvestris*. *Developmental Biology* 112:377–382.

Dermen, H. 1969. Directional cell division in shoot apices. *Cytologia* 34:541–558.

Donnison, I.S. and Francis, D. 1993. Determination of floral organ types in cultured *Silene* shoot apices. *Physiologia Plantarum* 89:315–322.

Donnison, I.S. and Francis, D. 1994. Experimental control of flower reversion on isolated shoot apices of the long day plant *Silene coeli-rosa*. *Physiologia Plantarum* 92:329–335.

Douady, S. and Couder, Y. 1992. Phyllotaxis as a physical self-organized growth process. *Physical Review Letters* 68:2098–2101.

Drews, G.N., Bowman, J.L. and Meyerowitz, E.M. 1991. Negative regulation of the *Arabidopsis* homeotic gene *AGAMOUS* by the *APETALA2* product. *Cell* 65:991–1002.

Dulieu, H. 1968. Emploi des chimères chlorophylliennes pour l'étude de l'ontogénie foliaire. *Bulletin de la Société de Bourgogne* 25:13–72.

Dulieu, H. 1969. Mutations somatiques chlorophylliennes et ontogénie caulinaire. *Bulletin de la Société de Bourgogne* 26:19–102.

Eames, A.J. and MacDaniels, L.H. 1951. *An Introduction to Plant Anatomy*. London: McGraw-Hill.

Edgar, E. 1961. *Fluctuations in Mitotic Index in the Shoot Apex of* Lonicera nitida. Christchurch: University of Canterbury.

Elliott, R.C., Betzner, A.S., Huttner, E., Oakes, M.P., Tucker, W.Q.J., Gerentes, D., Perez, P. and Smyth, D.R. 1996. *AINTEGUMENTA*, an *APETALA2*-like gene of *Arabidopsis* with pleiotropic roles in ovule development and floral organ growth. *Plant Cell* 8:155–168.

Erbar, C. 1988. Early developmental patterns in flowers and their value for systematics. In *Aspects of Floral Development*, eds P. Leins, S.C. Tucker and P.K. Endress, pp. 7–23. Berlin: J. Cramer.

Erickson, R.O. and Michelini, F.J. 1957. The plastochron index. *American Journal of Botany* 44:297–305.

Esau, K. 1965. *Plant Anatomy* (2nd Edition). New York: Wiley.

Evans, L.S. and Berg, A.R. 1971. Leaf and apical growth characteristics in *Triticum*. *American Journal of Botany* 58:540–543.

Evans, L.S. and Berg, A.R. 1972a. Qualitative histochemistry of the shoot apex of *Triticum*. *Canadian Journal of Botany* 50:241–244.

Evans, L.S. and Berg, A.R. 1972b. Early histogenesis and semiquantitative histochemistry of leaf initiation in *Triticum aestivum*. *American Journal of Botany* 59:973–980.

Evans, L.T. 1960. The influence of environmental conditions on inflorescence development in some long-day grasses. *New Phytologist* 59:163–174.

Evans, L.T. 1969. The nature of flower induction. In *The Induction of Flowering: Some Case Histories*, ed. L.T. Evans, pp. 457–480. Melbourne: Macmillan.

Evans, L.T., Knox, R.B. and Rijven, A.H.G.C. 1970. The nature and localization of early events in the shoot apex of *Lolium temulentum* during floral induction. In *Cellular and Molecular Aspects of Floral Induction*, ed. G. Bernier, pp. 192–206. London: Longmans.

Evans, P.T., Holaway, B.L. and Malmberg, R.L. 1988. Biochemical differentiation in the tobacco flower probed with monoclonal antibodies. *Planta* 175:259–269.

Evrard, J.-L., Jako, C., Saint-Guily, A., Weil, J.-H. and Kuntz, M. 1991. Anther-specific, developmentally regulated expression of genes encoding a new class of proline-rich proteins in sunflower. *Plant Molecular Biology* 16:271–281.

Felippe, G.M. and Dale, J.E. 1990. The effects of phosphate supply on growth of plants from the Brasilian Cerrado: experiments with seedlings of the annual weed, *Bidens gardneri* Baker (Compositeae) and the tree, *Qualea grandiflora* (Mart.) (Vochysiaceae). *Oecologia* 82:81–86.

Fernandez, D.E., Turner, F.R. and Crouch, M.L. 1991. *In situ* localization of storage protein nRNAs in developing meristems of *Brassica napus* embryos. *Development* 111:299–313.

Fleming, A.J. and Kuhlemeier, C. 1994. Activation of basal cells of the apical meristem during sepal formation in tomato. *Plant Cell* 6:789–798.

Fleming, A.J., Mandel, T., Hofmann, S., Sterk, P., De Vries, S.C. and Kuhlemeier, C. 1992. Expression pattern of a tobacco lipid transfer protein gene within the shoot apex. *Plant Journal* 2:855–862.

Fleming, A.J., Mandel, T., Roth, I. and Kuhlemeier, C. 1993. The patterns of gene expression in the tomato shoot apical meristem. *Plant Cell* 5:297–309.

Foard D.E., Haber, A.H. and Fishman, T.N. 1965. Initiation of lateral root primordia without completion of mitosis and without cytokinesis in uniseriate pericycle. *American Journal of Botany* 52:580–590.

Foard, D.E. 1971. The initial protrusion of a leaf primordium can form without concurrent periclinal cell divisions. *Canadian Journal of Botany* 49:1601–1603.

Fobert, P.R., Coen, E.S., Murphy, G.J.P. and Doonan, J.H. 1994. Patterns of cell division revealed by transcriptional regulation of genes during the cell cycle in plants. *EMBO Journal* 13:616–624.

Fosket, D.E. and Miksche, J.P. 1966. A histochemical study of the seedling shoot apical meristem of *Pinus lambertiana*. *American Journal of Botany* 53:694–702.

Fowler, M.W. and ap Rees, T. 1970. Carbohydrate oxidation during differentiation in roots of *Pisum sativum*. *Biochimica et Biophysica Acta* 201:33–44.

Francis, D. 1981a. A rapid accumulation of cells in G2 in the shoot apex of *Silene coeli-rosa* during the first day of floral induction. *Annals of Botany* 48:391–394.

Francis, D. 1981b. Effects of red and far-red light on cell division in the shoot apex of *Silene coeli-rosa* L. *Protoplasma* 107:285–299.

Francis, D. 1992. The cell cycle in plant development. *New Phytologist* 122:1–20.

Francis, D. 1997. The stem cell concept applied to shoot meristems of higher plants. In *Stem Cells*, ed. C.S. Potten, pp. 59–73. London: Academic Press.

Francis, D. and Barlow, P.W. 1988. Temperature and the cell cycle. In *Plants and Temperature* (42nd Symposium of the Society for Experimental Biology), eds S.P. Long and F.I. Woodward, pp. 181–201. Cambridge: Company of Biologists.

Francis, D. and Halford, N.G. 1995. The plant cell cycle. *Physiologia Plantarum* 93:365–374.

Francis, D. and Lyndon, R.F. 1978a. Early effects of floral induction on cell division in the shoot apex of *Silene*. *Planta* 139:273–279.

Francis, D. and Lyndon, R.F. 1978b. The cell cycle of the shoot apex of *Silene* during the first day of floral induction. *Protoplasma* 96:81–88.

Francis, D. and Lyndon, R.F. 1979. Synchronisation of cell division in the shoot apex of *Silene* in relation to flower initiation. *Planta* 145:151–157.

Francis, D. and Lyndon, R.F. 1985. The control of the cell cycle in relation to floral induction. In *The Cell Division Cycle in Plants*, eds J. Bryant and D. Francis, pp. 199–215. Cambridge: Cambridge University Press.

Francis, D., Rembur, J. and Nougarède, A. 1988. Changements dans la composition polypeptidique du méristème de *Silene coeli-rosa* (L.) au cours de l'induction florale. *C.R. Acad. Sci. Paris, Ser. III* 307:763–770.

Fujie, M., Kuroiwa, H., Kawano, S., Mutoh, S. and Kuroiwa, T. 1994. Behavior of organelles and their nucleoids in the shoot apical meristem during leaf development in *Arabidopsis thaliana* L. *Planta* 194:395–405.

Furuya, M. 1958. A biomorphological significance and an analysis of the effects of auxin on the development of 'composed' shoot in bean plants. *Journal of the Faculty of Science Tokyo III* 7:451–470.

Gahan, P.B. and Bellani, L.M. 1984. Identification of shoot apical meristem cells committed to form vascular elements in *Pisum sativum* L. and *Vicia faba* L. *Annals of Botany* 54:837–841.

Gasser, C.S. 1991. Molecular studies on the differentiation of floral organs. *Annual Review of Plant Physiology and Plant Molecular Biology* 42:621–649.

Gasser, C.S., Budelier, K.A., Smith, A.G., Shah, D.M. and Fraley, R.T. 1989. Isolation of tissue-specific cDNAs from tomato pistils. *Plant Cell* 1:15–24.

Gavaudan, P. and Gastelier, J. 1970. Organisation en groupes cellulaires de l'apex végétatif de *Phaseolus multiflorus* L. *Comptes Rendus de la Société de Biologie de Poitiers* 164:362–366.

Giddings, T.H. and Staehelin, L.A. 1991. Microtubule-mediated control of microfibril deposition: A re-examination of the hypothesis. In *The Cytoskeletal Basis of Plant Growth and Form*, ed. C.W. Lloyd, pp. 85–99. London: Academic Press.

Gifford, E.M. 1954. The shoot apex in angiosperms. *Botanical Review* 20:477–529.

Gifford, E.M. 1983. Concept of apical cells in bryophytes and pteridophytes. *Annual Review of Plant Physiology* 34:419–440.

Gifford, E.M. 1985. The apical cell of fern roots and shoots: an appraisal of its functional role in development. *Proceedings of the Royal Society of Edinburgh* 86B:237–243.

Gifford, E.M. and Corson, G.E. 1971. The shoot apex in seed plants. *Botanical Review* 37:143–229.

Gifford, E.M. and Kurth, E. 1983. Quantitative studies of the vegetative shoot apex of *Equisetum scirpoides*. *American Journal of Botany* 70:74–79.

Gifford, E.M. and Polito, V.S. 1981. Mitotic activity at the shoot apex of *Azolla filiculoides*. *American Journal of Botany* 68:1050–1055.

Gifford, E.M. and Stewart, K.D. 1967. Ultrastructure of the shoot apex of *Chenopodium album* and certain other seed plants. *Journal of Cell Biology* 33:131–142.

Gifford, E.M. and Tepper, H.B. 1962. Ontogenetic and histochemical changes in the vegetative shoot tip of *Chenopodium album*. *American Journal of Botany* 49:902–911.

Goldberg, R.B. 1986. Regulation of plant gene expression. *Philosophical Transactions of the Royal of Society of London B* 314:343–353.

Goldsmith, M.H.M. 1977. The polar transport of auxin. *Annual Review of Plant Physiology* 28:439–478.

Golub, S.J. and Wetmore, R.H. 1948. Studies of development in the vegetative shoot of *Equisetum arvense* L. I. The shoot apex. *American Journal of Botany* 35:755–767.

Gonthier, R. and Francis, D. 1989. Changes in the pattern of cell division in the shoot and root meristems of *Lolium perenne* during the transition from vegetative to floral growth. *Journal of Experimental Botany* 40:285–292.

Gonthier, R., Jacqmard, A. and Bernier, G. 1985. Occurrence of two cell sub-populations with different cell-cycle durations in the central and peripheral zones of the vegetative shoot apex of *Sinapis alba* L. *Planta* 165:288–291.

Gonthier, R., Jacqmard, A. and Bernier, G. 1987. Changes in cell cycle duration and growth fraction in the shoot meristem of *Sinapis* during floral induction. *Planta* 170:55–59.

Goodwin, B. 1995. *How the Leopard Changed its Spots*. London: Phoenix.

Goodwin, B.C. and Saunders, P.T. (eds) 1989. *Theoretical Biology: Epigenetic and Evolutionary Order from Complex Systems*. Edinburgh: Edinburgh University Press.

Goodwin, P.B. 1978. Phytohormones and growth and development of organs of the vegetative plant. In *Phytohormones and Related Compounds*, vol. 2, eds D.S. Letham, P.B. Goodwin and T.J.V. Higgins, pp. 31–173. Amsterdam: Elsevier.

Goodwin, P.B. and Lyndon, R.F. 1983. Synchronisation of cell division during transition to flowering in *Silene* apices not due to increased symplast permeability. *Protoplasma* 116:219–222.

Goto, K. and Meyerowitz, E.M. 1994. Function and regulation of the *Arabidopsis* floral homeotic gene *PISTILLATA*. *Genes and Development* 8:1548–1560.

Grayburn, W.S., Green, P.B. and Steucek, G. 1982. Bud induction with cytokinin. A local response to local application. *Plant Physiology* 69:682–686.

Green, P.B. 1974. Morphogenesis of the cell and organ axis – biophysical models. *Brookhaven Symposia in Biology* 25:166–190.

Green, P.B. 1976. Growth and cell pattern formation on an axis: critique of concepts, terminology and modes of study. *Botanical Gazette* 137:187–202.

Green, P.B. 1980. Organogenesis – a biophysical view. *Annual Review of Plant Physiology* 31:51–82.

Green, P.B. 1984. Analysis of axis extension. In *Positional Controls in Plant Development*, eds P.W. Barlow and D.J. Carr, pp. 53–82. Cambridge: Cambridge University Press.

Green, P.B. 1985. Surface of the shoot apex: a reinforcement-field theory for phyllotaxis. *Journal of Cell Science Supplement* 2:181–201.

Green P.B. 1986. Plasticity in shoot development: a biophysical view. In *Plasticity in Plants* (Symposia of the Society for Experimental Biology, 40), eds D.H.

Jennings and A.J. Trewavas, pp. 211–232. Cambridge: Company of Biologists.

Green, P.B. 1988. A theory for inflorescence development and flower formation based on morphological and biophysical analysis in *Echeveria*. *Planta* 175:153–169.

Green, P.B. 1994. Connecting gene and hormone action to form, pattern and organogenesis: biophysical transductions. *Journal of Experimental Botany* 45:1775–1788.

Green, P.B. 1996. Expression of form and pattern in plants – a role for biophysical fields. *Cell and Developmental Biology* 7:903–911.

Green, P.B. and Brooks, K.E. 1978. Stem formation from a succulent leaf: its bearing on theories of axiation. *American Journal of Botany* 65:13–26.

Green, P.B., Havelange, A. and Bernier, G. 1991. Floral morphogenesis in *Anagallis*: scanning-electron-micrograph sequences from individual growing meristems before, during and after the transition to flowering. *Planta* 185:502–512.

Green, P.B. and Poethig, R.S. 1982. Biophysics of the extension and initiation of plant organs. In *Developmental Order: Its Origin and Regulation*, eds S. Subtelny and P.B. Green, pp. 485–509. New York: Alan R. Liss.

Green, P.B. and Selker, J.M.L. 1991. Mutual alignments of cell walls, cellulose and cytoskeletons: their role in meristems. In *The Cytoskeletal Basis of Plant Growth and Form*, ed. C. Lloyd, pp. 303–322. London: Academic Press.

Green, P.B., Steele, C.S. and Rennich, S.C. 1996. Phyllotactic patterns: a biophysical mechanism for their origin. *Annals of Botany* 77:515–527.

Gregory, R.A. and Romberger, J.A. 1972a. The shoot apical ontogeny of the *Picea abies* seedling. I. Anatomy, apical dome diameter and plastochron duration. *American Journal of Botany* 59:587–597.

Gregory, R.A. and Romberger, J.A. 1972b. The shoot apical ontogeny of the *Picea abies* seedling. II. Growth rates. *American Journal of Botany* 59:598–606.

Greyson, R.I. 1994. *The Development of Flowers*. New York: Oxford University Press.

Greyson, R.I. and Walden, D.B. 1972. The ABPHYL syndrome in *Zea mays*. I. Arrangement, number and size of leaves. *American Journal of Botany* 59:466–472.

Greyson, R.I., Walden, D.B., Humes, J.A. and Erickson, R.O. 1978. The ABPHYL syndrome in *Zea mays*. II. Patterns of leaf initiation and the shape of the shoot meristem. *Canadian Journal of Botany* 56:1545–1550.

Griffiths, F.E.W., Lyndon, R.F. and Bennett, M.D. 1985. The effects of vernalization on the growth of the wheat shoot apex. *Annals of Botany* 56:501–511.

Griffiths, F.W. 1981. The effects of vernalisation on growth at the wheat shoot apex. MPhil thesis, University of Edinburgh.

Grose, S. and Lyndon, R.F. 1984. Inhibition of growth and synchronised cell division in the shoot apex in relation to flowering in *Silene*. *Planta* 161:289–294.

Gunning, B.E.S., Hughes, J.E. and Hardham, A.R. 1978. Formative and proliferative cell divisions, cell differentiation and developmental changes in the meristem of *Azolla* roots. *Planta* 143:121–144.

Gunning, B.E.S. and Hardham, A.R. 1982. Microtubules. *Annual Review of Plant Physiology* 33:651–698.

Hackett, W.P., Cordero, R.E. and Srinivasan, C. 1987. Apical meristem characteristics and activity in relation to juvenility in *Hedera*. In *Manipulation of Flowering*, ed. J.G. Atherton, pp. 93–99. London: Butterworths.

Haley, A., Russell, A.J., Wood, N., Allan, A.C., Knight, M., Campbell, A.K. and Trewavas, A.J. 1995. Effects of mechanical signalling on plant cytosolic calcium. *Proceedings of the National Academy of Sciences, USA* 92:4124–4128.

Hallet, J.-N. 1969. Remarques sur la structure et sur le fonctionnement de l'apex du *Polytrichum formosum* Hedw. *C.R. Acad. Sci. Paris D* 268:916–919.

Hallet, J.-N. 1974. Morphogenèse du gamétophyte feuillé du *Polytrichum formosum* Hedw. II. Etude histochimique, histoautoradiographique et cytophotométriqe de la région apicale pendant la phase reproductrice. *Annales des Sciences Naturelles, Botanique, Paris* 15:321–388.

Hallet, J.-N. 1976. Les cycle mitotiques dans le point végétatif de l'*Hookeria lucens* (Hedw.) SM. (Mousse pleurocarpe, Hookeriale). *Journal of Bryology* 9:97–103.

Hallet, J.-N. 1977. Le cycle cellulaire de l'apicale muscinale: données nouvelles et caractères originaux. *Bryophytorum Bibliotheca* 13:1–20.

Hallet, J.-N. 1978. RNA synthesis in the shoot apex of *Polytrichum formosum* Hedw. *Annals of Botany* 42:381–387.

Hantke, S.S., Carpenter, R. and Coen, E.S. 1995. Expression of *floricaula* in single cell layers of periclinal chimeras activates downstream homeotic genes in all layers of floral meristems. *Development* 121:27–35.

Hara, N. 1971. Structure of the vegetative shoot apex of *Clethra barbinervis*. I. Superficial and transectional views. *Botanical Magazine Tokyo* 84:8–17.

Hara, N. 1980. Morphological study on early ontogeny of the *Ginkgo* leaf. *Botanical Magazine Tokyo* 93:1–12.

Hardenack, S., Ye, D., Saedler, H. and Grant, S. 1994. Comparison of MADS box gene expression in developing male and female flowers of the dioecious plant white campion. *Plant Cell* 6:1775–1787.

Harte, C. and Lindenmayer, A. 1983. Mitotic index in growing cell populations: mathematical models and computer simulations. *Biologisches Zentralblatt* 102:509–533.

Havelange, A. 1980. The quantitative ultrastructure of the meristematic cells of *Xanthium strumarium* during the transition to flowering. *American Journal of Botany* 67:1171–1178.

Havelange, A. and Bernier, G. 1983. Partial floral evocation by high irradiance in the long-day plant *Sinapis alba*. *Physiologia Plantarum* 59:545–550.

Havelange, A., Bernier, G. and Jacqmard, A. 1974. Descriptive and quantitative study of ultrastructural changes in the apical meristem of mustard in transition to flowering. II. The cytoplasm, mitochondria and proplastids. *Journal of Cell Science* 16:421–432.

Havelange, A., Bodson, M. and Bernier, G. 1986. Partial floral evocation by exogenous cytokinin in the long-day plant *Sinapis alba*. *Physiologia Plantarum* 67:695–701.

Hay, R.K.M. and Kemp, D.R. 1990. Primordium initiation at the stem apex as the primary event controlling plant development: preliminary evidence from wheat for the regulation of leaf development. *Plant Cell and Environment* 13:1005–1008.

Hébant, C., Hébant-Mauri, R. and Barthonnet, J. 1978. Evidence for division and polarity in apical cells of bryophytes and pteridophytes. *Planta* 138:49–52.

Hébant-Mauri, R. 1975. Apical segmentation and leaf initiation in the tree fern *Dicksonia squarrosa*. *Canadian Journal of Botany* 53:764–772.

Hejnowicz, Z. 1955. Growth distribution and cell arrangement in apical meristems. *Acta Societatis Botanicorum Poloniae* 24:583–608.

Hejnowicz, Z. 1982. Vector and scalar fields in modeling of spatial variations of growth rates within plant organs. *Journal of Theoretical Biology* 96:161–173.

Hejnowicz, Z. 1984. Trajectories of principal directions of growth, natural coordinate system in growing plant organ. *Acta Societatis Botanicorum Poloniae* 53:29–42.

Hejnowicz, Z. and Nakielski, J. 1979. Modeling of growth in shoot apical dome. *Acta Societatis Botanicorum Poloniae* 48:423–441.

Hejnowicz, Z., Nakielski, J. and Hejnowicz, K. 1984a. Modeling of spatial variations of growth within apical domes by means of the growth tensor. I. Growth specified on dome axis. *Acta Societatis Botanicorum Poloniae* 53:17–28.

Hejnowicz, Z., Nakielski, J. and Hejnowicz, K. 1984b. Modeling of spatial variations of growth within apical domes by means of the growth tensor. II. Growth specified on dome surface. *Acta Societatis Botanicorum Poloniae* 53:301–316.

Herbert, R.J., Francis, D. and Ormrod, J.C. 1992. Cellular and morphological changes at the terminal shoot apex of the short day plant, *Pharbitis nil* (*Ipomoea nil* L.) cv violet, during the transition to flowering. *Physiologia Plantarum* 86:85–92.

Hernández, L.F. and Green, P.B. 1993. Transductions for the expression of structural pattern: analysis in sunflower. *Plant Cell* 5:1725–1738.

Hernández, L.F. and Palmer, J.H. 1988. Regeneration of the sunflower capitulum after cylindrical wounding of the receptacle. *American Journal of Botany* 75:1253–1261.

Heslop-Harrison, J. and Heslop-Harrison, Y. 1970. The state of the apex and the response to induction in *Cannabis sativa*. In *Cellular and Molecular Aspects of Floral Induction*, ed. G. Bernier, pp. 3–26. London: Longman.

Hicks, G.S. 1980. Patterns of organ development in plant tissue culture and the problem of organ determination. *Botanical Review* 46:1–23.

Houssa, C., Bernier, G., Pieltain, A., Kinet, J.-M. and Jacqmard, A. 1994. Activation of latent DNA-replication origins: a universal effect of cytokinins. *Planta* 193:247–250.

Houssa, C., Jacqmard, A. and Bernier, G. 1990. Activation of replicon origins as a possible target for cytokinins in shoot meristems of *Sinapis*. *Planta* 181:324–326.

Huang, J., Struck, F., Matzinger, D.F. and Levings, C.S. 1994. Flower-enhanced expression of a nuclear-encoded mitochondrial respiratory protein is associated with changes in mitochondrion number. *Plant Cell* 6:439–448.

Huijser, P., Klein, J., Lönnig, W.-E., Meijer, H., Saedler, H. and Sommer, H. 1992. Bracteomania, an inflorescence anomaly, is caused by loss of function of the MADS-box gene *squamosa* in *Antirrhinum majus*. *EMBO Journal* 11:1239–1249.

Hussey, G. 1971a. *In vitro* growth of vegetative tomato shoot apices. *Journal of Experimental Botany* 22:688–701.

Hussey, G. 1971b. Cell division and expansion and resultant tissue tensions in the shoot apex during the formation of a leaf primordium in the tomato. *Journal of Experimental Botany* 22:702–714.

Hussey, G. 1972. The mode of origin of a leaf primordium in the shoot apex of the pea (*Pisum sativum*). *Journal of Experimental Botany* 23:675–682.

Hussey, G. 1973. Mechanical stress in the shoot apices of *Euphorbia*, *Lycopersicon* and *Pisum* under controlled turgor. *Annals of Botany* 37:57–64.

Irish, V. 1991. Cell lineage in plant development. *Current Opinion in Genetics and Development* 1:169–173.

Irish, V.F. and Sussex, I.M. 1990. Function of the *apetala-1* gene during *Arabidopsis* floral development. *Plant Cell* 2:741–753.

Iskander, S.R. and Brossard-Chriqui, D. 1980. Potentialités organogènes des disques foliaires du *Datura innoxia* Mill. cultivés *in vitro*. *Zeitschrift für Pflanzenphysiologie* 98:245–254.

Jackson, D., Roberts, K. and Martin, C. 1992. Temporal and spatial control of expression of anthocyanin biosynthetic genes in developing flowers of *Antirrhinum majus*. *Plant Journal* 2:425–434.

Jackson, D., Veit, B. and Hake, S. 1994. Expression of maize *knotted1* related homeobox genes in the shoot apical meristem predicts patterns of morphogenesis in the vegetative shoot. *Development* 120:405–413.

Jackson, J.A. and Lyndon, R.F. 1990. Habituation: cultural curiosity or developmental determinant? *Physiologia Plantarum* 79:579–583.

Jacobs, B.C. and Pearson, C.J. 1992. Pre-flowering growth and development of the inflorescences of maize. I. Primordia production and apical dome volume. *Journal of Experimental Botany* 43:557–563.

Jacqmard, A. 1968. Early effects of gibberellic acid on mitotic activity and DNA synthesis in the apical bud of *Rudbeckia bicolor*. *Physiologie Végétale* 6:409–416.

Jacqmard, A. 1970. Duration of the mitotic cycle in the apical bud of *Rudbeckia bicolor*. *New Phytologist* 69:269–271.

Jacqmard, A. 1978. Histochemical localization of enzyme activity during floral evocation in the shoot apical meristem of *Sinapis alba*. *Protoplasma* 94:315–324.

Jacqmard, A., Bomans, J., Ormrod, J.C. and Bernier, G. 1993. Early increase in the mitotic index in the shoot apex of *Lolium temulentum* cv. Ceres during the floral transition. *Journal of Experimental Botany* 44:1407–1409.

Jacqmard, A. and Houssa, C. 1988. DNA fiber replication during a morphogenetic switch in the shoot meristematic cells of a higher plant. *Experimental Cell Research* 179:454–461.

Jacqmard, A., Houssa, C. and Bernier, G. 1994. Regulation of the cell cycle by cytokinins. In *Cytokinins: Chemistry, Activity and Function*, eds. D.W.S. Mok and M.C. Mok, pp. 197–215. Boca Raton: CRC Press.

Jacqmard, A., Houssa, C. and Bernier, G. 1995. Abscisic acid antagonises the effect of cytokinin on DNA-replication origins. *Journal of Experimental Botany* 46:663–666.

Jacqmard, A., Lyndon, R.F. and Salmon, J. 1984. Appearance of specific antigenic proteins in the maturing sexual organs of *Sinapis* flowers. *Journal of Cell Science* 68:195–209.

Jacqmard, A. and Miksche, J. 1971. Cell population and quantitative changes of DNA in the shoot apex of *Sinapis alba* during floral induction. *Botanical Gazette* 132:364–367.

Jacqmard, A., Raju, M.V.S., Kinet, J.-M. and Bernier, G. 1976. The early action of the floral stimulus on mitotic activity and DNA synthesis in the apical meristem of *Xanthium strumarium*. *American Journal of Botany* 63:166–174.

Jean, R.V. 1994. *Phyllotaxis. A Systematic Study in Plant Morphogenesis*. New York: Cambridge University Press.

Jegla, D.E. and Sussex, I.M. 1989. Cell lineage patterns in the shoot meristem of the sunflower embryo in the dry seed. *Developmental Biology* 131:215–225.

Jensen, L.C.W. 1971. Experimental bisection of *Aquilegia* floral buds cultured *in vitro*. I. The effect on growth, primordia initiation and apical regeneration. *Canadian Journal of Botany* 49:487–493.

Jernstedt, J.A., Cutter, E.G. and Lu, P. 1994. Independence of organogenesis and cell pattern in developing angle shoots of *Selaginella martensii*. *Annals of Botany* 74:343–355.

Jesuthasan, S. and Green, P.B. 1989. On the mechanism of decussate phyllotaxis: biophysical studies on the tunica layer of *Vinca major*. *American Journal of Botany* 76:1152–1166.

Kanchanapoom, M.L. and Thomas, J.F. 1987a. Stereological study of ultrastructural changes in the shoot apical meristem of *Nicotiana tabacum* during floral induction. *American Journal of Botany* 74:152–163.

Kanchanapoom, M.L. and Thomas, J.F. 1987b. Quantitative ultrastructural changes in tunica and corpus cells of the shoot apex of *Nicotiana tabacum* during transition to flowering. *American Journal of Botany* 74:241–249.

Kaplan, D.R. 1970. Comparative foliar histogenesis in *Acorus calamus* and its bearing on the phyllode theory of monocotyledonous leaves. *American Journal of Botany* 57:331–361.

Kelly, A.J., Zagotta, M.T., White, R.A., Chang, C. and Weeks-Wagner, D.R. 1990. Identification of genes expressed in the tobacco shoot apex during the floral transition. *Plant Cell* 2:963–972.

Kinet, J.-M., Bodson, M., Alvinia, A.M. and Bernier, G. 1971. The inhibition of flowering in *Sinapis alba* after the arrival of the floral stimulus at the meristem. *Zeitschrift für Pflanzenphysiologie* 66:49–63.

Kinet, J.-M., Sachs, R.M. and Bernier, G. 1985. *The Physiology of Flowering*, vol. III. Boca Raton: CRC Press.

Kinsman, E.A., Lewis, C., Davies, M.S., Young, J.E., Francis, D., Thomas, I.D. and Chorlton, K.H. 1996. Effects of temperature and elevated CO_2 on cell division in shoot meristems: differential responses of two natural populations of *Dactylis glomerata* L. *Plant, Cell and Environment* 19:775–780.

Kirby, E.J.M. 1974. Ear development in spring wheat. *Journal of Agricultural Science* 82:437–447.

Kirk, J.T.O. and Tilney-Bassett, R.A.E. 1978. *The Plastids: Their Chemistry, Structure, Growth and Inheritance* (2nd Edition). Amsterdam: Elsevier.

Klopfer, A. von, 1973. Histochemische Untersuchungen über die Enzymaktivität in vegetativen und floralen Sprossscheiteln. 1. Zytochromoxydase. *Biochemie und Physiologie der Pflanzen* 164:383–396.

Klucher, K.M., Chow, H., Reiser, L. and Fischer, R.L. 1996. The *AINTEGUMENTA* gene of *Arabidopsis* required for ovule and female gametophyte development is related to the floral homeotic gene *APETALA2*. *Plant Cell* 8:137–153.

Knott, J.E. 1934. Effect of a localized photoperiod on spinach. *Proceedings of the American Society for Horticultural Science* 31:152–154.

Knox, R.B. and Evans, L.T. 1966. Inflorescence initiation in *Lolium temulentum* L. VIII. Histochemical changes at the shoot apex during induction. *Australian Journal of Biological Sciences* 19:233–245.

Köhler, S., Coraggio, I., Becker, D. and Salamini, F. 1992. Pattern of expression of meristem-specific cDNA clones of barley (*Hordeum vulgare* L.). *Planta* 186:227–235.

Koltunow, A.M., Truettner, J., Cox, K.N., Wallroth, M. and Goldberg, R.B. 1990. Different temporal and spatial gene expression patterns occur during anther development. *Plant Cell* 2:1201–1224.

Komaki, M.K., Okada, K., Nishino, E. and Shimura, Y. 1988. Isolation and characterization of novel mutants of *Arabidopsis thaliana* defective in flower development. *Development* 104:195–203.

Konig, A.J., Tanimoto, E.Y., Kiehne, K., Rost, T. and Comai, L. 1991. Cell-specific expression of plant histone H2A genes. *Plant Cell* 3:657–665.

Körner, C. and Menendez-Riedl, S. 1989. The significance of developmental aspects in plant growth analysis. In *Causes and Consequences of Variation in Growth Rate and Productivity of Higher Plants*, ed. H. Lambers, pp. 141–157. The Hague: Academic Publishing.

Kosugi, S., Suzuka, I., Ohashi, Y., Murakami, T. and Arai, Y. 1991. Upstream sequences of rice proliferating cell nuclear antigen (PCNA) gene mediate expression of PCNA-GUS chimeric gene in meristems of transgenic tobacco plants. *Nucleic Acids Research* 19:1571–1576.

Krekule, J. and Teltscherova, L. 1966. Mikrorespirometrische Bestimmung der Atmungsintensität an Vegetationskegeln von Weizenpflanzen beim Übergang von der vetativen zur generativen Entwicklungsphase. *Biologia Plantarum* 8:299–304.

Kuehnert, C.C. and Miksche, J.P. 1964. Application of the 22.5 MEV deuteron microbeam to the study of morphogenetic problems within the shoot apex of *Osmunda claytonia*. *American Journal of Botany* 51:743–747.

Kunst, L., Klenz, J.E., Martinez-Zapater, J. and Haughn, G.W. 1989. *Ap 2* gene determines the identity of perianth organs in flowers of *Arabidopsis thaliana*. *Plant Cell* 1:1195–1208.

Kutschera, U. 1992. The role of the epidermis in the control of elongation growth in stems and coleoptiles. *Botanica Acta* 105:195–203.

Kutschera, U., Bergfeld, R. and Schopfer, P. 1987. Cooperation of epidermis and inner tissues in auxin-mediated growth of maize coleoptiles. *Planta* 170:168–180.

Kutschera, U. and Briggs, W.R. 1987. Differential effect of auxin on the *in vivo* extensibility of cortical cylinder and epidermis in pea internodes. *Plant Physiology* 84:1361–1366.

Langenauer, H.D., Davis, E.L. and Webster, P.L. 1974. Quiescent cell populations in apical meristems of *Helianthus annuus*. *Canadian Journal of Botany* 52:2195–2201.

Larson, P.R. 1983. Primary vascularization and the siting of primordia. In *The Growth and Functioning of Leaves*, eds J.E. Dale and F.L. Milthorpe, pp. 25–51. Cambridge: Cambridge University Press.

Law, C.N., Worland, A.J. and Giorgi, B. 1976. The genetic control of ear-emergence time by chromosomes 5A and 5D of wheat. *Heredity* 36:49–58.

Leins, P., Tucker, S.C. and Endress, P.K. (eds) 1988. *Aspects of Floral Development*. Berlin: J. Cramer.

Leshem, B. and Clowes, F.A.L. 1972. Rates of mitosis in shoot apices of potatoes at the beginning and end of dormancy. *Annals of Botany* 36:687–691.

Leyser, H.M.O. and Furner, I.J. 1992. Characterisation of three shoot apical meristem mutants of *Arabidopsis thaliana*. *Development* 116:397–403.

Lin, J. and Gifford, E.M. 1976. The distribution of ribosomes in the vegetative and floral apices of *Adonis aestivalis*. *Canadian Journal of Botany* 54:2478–2483.

Lincoln, C., Long, J., Yamaguchi, T., Serikawa, K. and Hake, S. 1994. A *knotted1*-like homeobox gene in *Arabidopsis* is expressed in the vegetative meristem and dramatically alters leaf morphology when overexpressed in transgenic plants. *Plant Cell* 6:1859–1876.

Lintilhac, P.M. 1974. Differentiation, organogenesis and the tectonics of cell wall orientation. III. Theoretical considerations of cell wall mechanics. *American Journal of Botany* 61:230–237.

Lintilhac, P.M. 1984. Positional controls in meristem development: a caveat and an alternative. In *Positional Controls in Plant Development*, eds P.W. Barlow and D.J. Carr, pp. 83–105. Cambridge: Cambridge University Press.

Lintilhac, P.M. and Green, P.B. 1976. Patterns of microfibrillar order in a dormant fern apex. *American Journal of Botany* 63:726–728.

Liu, Z.C., Running, M.P. and Meyerowitz, E.M. 1997. *TSO1* functions in cell division during *Arabidopsis* flower development. *Development* 124:665–672.

Lloyd, C. 1995. Life on a different plane. *Current Biology* 5:1085–1087.

Loiseau, J.-E. 1962. Activité mitotique des cellules superficielles du sommet végétatif caulinaire. *Bulletin de la Société Botanique de France, Mémoires* 14–23.

Loiseau, J.E. 1969. *La Phyllotaxie*. Paris: Masson.

Lord, E.M. 1991. The concepts of heterochrony and homeosis in the study of floral morphogenesis. *Flowering Newsletter* 11:4–13.

Lynch, D.V. and Rivera, E.R. 1981. Ultrastructure of cells in the overwintering dormant shoot apex of *Rhododendron maximum* L. *Botanical Gazette* 142:63–72.

Lyndon, R.F. 1967. The growth of the nucleus in dividing and non-dividing cells of the pea root. *Annals of Botany* 31:133–146.

Lyndon, R.F. 1968a. Changes in volume and cell number in the different regions of the shoot apex of *Pisum* during a single plastochron. *Annals of Botany* 32:371–390.

Lyndon, R.F. 1968b. The structure, function and development of the nucleus. In *Plant Cell Organelles*, ed. J. Pridham, pp. 16–39. London: Academic Press.

Lyndon, R.F. 1970a. Rates of cell division in the shoot apical meristem of *Pisum*. *Annals of Botany* 34:1–17.

Lyndon, R.F. 1970b. Planes of cell division and growth in the shoot apex of *Pisum*. *Annals of Botany* 34:19–28.

Lyndon, R.F. 1970c. DNA, RNA and protein in the pea shoot apex in relation to leaf initiation. *Journal of Experimental Botany* 21:286–291.

Lyndon, R.F. 1971. Growth of the surface and inner parts of the pea shoot apical meristem during leaf initiation. *Annals of Botany* 35:263–270.

Lyndon, R.F. 1972a. Nucleic acid synthesis in the pea shoot apex. *Symposia Biologica Hungarica* 13:345–353.

Lyndon, R.F. 1972b. Leaf formation and growth at the shoot apical meristem. *Physiologie Végétale* 10:209–222.

Lyndon, R.F. 1973. The cell cycle in the shoot apex. In *The Cell Cycle in Development and Differentiation*, eds M. Balls and F.S. Billett, pp. 167–183. Cambridge: Cambridge University Press.

Lyndon, R.F. 1976. The shoot apex. In *Cell Division in Higher Plants*, ed. M.M. Yeoman, pp. 285–314. London: Academic Press.

Lyndon, R.F. 1977a. The shoot apical meristem. In *The Physiology of the Garden Pea*, eds J. Pate and J.F. Sutcliffe, pp. 183–211. London: Academic Press.

Lyndon, R.F. 1977b. Interacting processes in development at the shoot apex. *Symposia of the Society for Experimental Biology* 31:221–250.

Lyndon, R.F. 1978a. Flower development in *Silene*: morphology and sequence of initiation of primordia. *Annals of Botany* 42:1343–1348.

Lyndon, R.F. 1978b. Phyllotaxis and the initiation of primordia during flower development in *Silene*. *Annals of Botany* 42:1349–1360.

Lyndon, R.F. 1979a. Rates of growth and primordial initiation during flower development in *Silene* at different temperatures. *Annals of Botany* 43:539–551.

Lyndon, R.F. 1979b. The cellular basis of apical differentiation. In *Control of Plant Development*, ed E.C. George, (British Plant Growth Regulator Group, Monograph 3), pp. 57–73.

Lyndon, R.F. 1982. Changes in polarity of growth during leaf initiation in the pea, *Pisum sativum* L. *Annals of Botany* 49:281–290.

Lyndon, R.F. 1987a. Initiation and growth of internodes and stem and flower frusta in *Silene coeli-rosa*. In *The Manipulation of Flowering*, ed. J. Atherton, pp. 301–314. London: Butterworths.

Lyndon, R.F. 1987b. Synchronization of cell division during flower initiation in third-order buds of *Silene*. *Annals of Botany* 59:67–72.

Lyndon, R.F. 1990. *Plant Development: The Cellular Basis*. London: Unwin Hyman.

Lyndon, R.F. 1992. Environmental control of reproductive development. In *Environmental Physiology and Ecology of Fruit and Seed Production*, eds C. Marshall and J. Grace, pp. 9–32. Cambridge: Cambridge University Press.

Lyndon, R.F. 1994. Control of organogenesis at the shoot apex. *New Phytologist* 128:1–18.

Lyndon, R.F. 1998. Phyllotaxis in the flower and in flower reversion. In *Symmetry in Plants*, ed. R.V. Jean. Singapore: World Scientific Press. [in press]

Lyndon, R.F. and Battey N.H. 1985. The growth of the shoot apical meristem during flower initiation. *Biologia Plantarum* 27:339–349.

Lyndon, R.F. and Cunninghame, M.E. 1986. Control of shoot apical development via cell division. In *Plasticity in Plants* (Symposia of the Society for Experimental Biology 40), eds D.H. Jennings and A.J. Trewavas, pp. 233–255. Cambridge: Company of Biologists

Lyndon, R.F. and Francis, D. 1984. The response of the shoot apex to light generated signals from the leaves. In *Light and the Flowering Process*, eds D. Vince-Prue, B. Thomas and K.E. Cockshull, pp. 171–189. London: Academic Press.

Lyndon, R.F. and Robertson, E.S. 1976. The quantitative ultrastructure of the pea shoot apex in relation to leaf initiation. *Protoplasma* 87:387–402.

Lyndon, R.F., Jacqmard, A. and Bernier, G. 1983. Changes in protein composition of the shoot meristem during floral evocation in *Sinapis alba*. *Physiologia Plantarum* 59:476–480.

Maksymowych, R. 1990. *Analysis of Growth and Development of Xanthium*. Cambridge: Cambridge University Press.

Maksymowych, R., Cordero, R.E. and Erickson, R.O. 1976. Long-term developmental changes in *Xanthium* induced by gibberellic acid. *American Journal of Botany* 63:1047–1053.

Maksymowych, R. and Erickson, R.O. 1977. Phyllotactic change induced by gibberellic acid in *Xanthium* shoot apices. *American Journal of Botany* 64:33–44.

Malmberg, R.L., McIndoo, J., Hiatt, J. and Lowe, B.A. 1985. Genetics of polyamine synthesis in tobacco: genetic switches in the flower. *Cold Spring Harbor Symposia in Quantitative Biology* 50:475–482.

Mandel, M.A., Bowman, J.L., Kempin, S.A., Ma, H., Meyerowitz, E.M. and Yanofsky, M.F. 1992. Manipulation of flower structure in transgenic tobacco. *Cell* 71:133–143.

Marc, J. and Hackett, W.P. 1989. A new method for immunofluorescent localization of microtubules in surface cell layers: application to the shoot apical meristem of *Hedera*. *Protoplasma* 148:70–79.

Marc, J. and Hackett, W.P. 1992. Changes in the pattern of cell arrangement at the surface of the shoot apical meristem in *Hedera helix* L. following gibberellin treatment. *Planta* 186:503–510.

Marc, J. and Palmer, H. 1984. Variation in cell-cycle time and nuclear DNA content in the apical meristem of *Helianthus annuus* L. during the transition to flowering. *American Journal of Botany* 71:588–595.

Marcotrigiano, M. 1986. Experimentally synthesised plant chimeras. 3. Qualitative and quantitative characteristics of the flowers of interspecific *Nicotiana* chimeras. *Annals of Botany* 57:435–442.

Martin, C., Prescott, A., Mackay, S., Bartlett, J. and Vrijlandt, E. 1991. Control of anthocyanin biosynthesis in flowers of *Antirrhinum majus*. *Plant Journal* 1:37–49.

Marushige, K. and Marushige, Y. 1962. An electrophoretic study of tissue extracts from leaf and flower in *Pharbitis nil* Chois. *Plant and Cell Physiology* 3:319–322.

Marx, G.A. 1977. A genetic syndrome affecting leaf development in *Pisum*. *American Journal of Botany* 64:273–277.

Masuda, Y. 1990. Auxin-induced cell elongation and cell wall changes. *Botanical Magazine Tokyo* 103:345–370.

Mauseth, J.D. 1976. Cytokinin- and gibberellic acid-induced effects on the structure and metabolism of shoot apical meristems in *Opuntia polycantha* (Cactaceae). *American Journal of Botany* 63:1295–1301.

Mauseth, J.D. 1980. A morphometric study of the ultrastructure of *Echinocereus engelmannii* (Cactaceae). I. Shoot apical meristems at germination. *American Journal of Botany* 67:173–181.

Mauseth, J.D. 1981a. A morphometric study of the ultrastructure of *Echinocereus engelmannii* (Cactaceae). II. The mature, zonate shoot apical meristem. *American Journal of Botany* 68:96–100.

Mauseth, J.D. 1981b. A morphometric study of the ultrastructure of *Echinocereus engelmannii* (Cactaceae). III. Subapical and mature tissues. *American Journal of Botany* 68:531–534.

Mauseth, J.D. 1982a. A morphometric study of the ultrastructure of *Echinocereus engelmannii* (Cactaceae). VI. The individualized ultrastructures of diverse types of meristems. *American Journal of Botany* 69:1524–1526.

Mauseth, J.D. 1982b. A morphometric study of the ultrastructure of *Echinocereus engelmannii* (Cactaceae). V. Comparison with the shoot apical meristems of *Trichocereus pachanoi* (Cactaceae). *American Journal of Botany* 69:551–555.

Mauseth, J.D. 1982c. A morphometric study of the ultrastructure of *Echinocereus engelmannii* (Cactaceae). IV. Leaf and spine primordia. *American Journal of Botany* 69:546–550.

McAlpin, B.W. and White, R.A. 1974. Shoot organization in the Filicales: the promeristem. *American Journal of Botany* 61:562–579.

McDaniel, C.N. 1984. Competence, determination and induction in plant development. In *Pattern Formation: A Primer in Developmental Biology*, ed. G.M. Malacinski, pp. 393–412. New York: Macmillan.

McDaniel, C.N. 1996. Developmental physiology of floral initiation in *Nicotiana tabacum* L. *Journal of Experimental Botany* 47:465–475.

McDaniel, C.N. and Poethig, R.S. 1988. Cell-lineage patterns in the shoot apical meristem of the germinating maize embryo. *Planta* 175:13–22.

McDaniel, C.N., Singer, S.R., Dennin, K.A. and Gebhardt, J.S. 1985. Floral determination: timing, stability and root influence. In *Plant Genetics* (UCLA Symposium on Cellular and Molecular Biology 35), ed. M. Freeling, pp. 73–87. New York: Alan R. Liss.

McDaniel, C.N., Singer, S.R. and Smith, S.M.E. 1992. Developmental states associated with the floral transition. *Developmental Biology* 153:59–69.

Medford, J.I. 1992. Vegetative apical meristems. *Plant Cell* 4:1029–1039.

Medford, J.I., Behringer, F.J., Callos, J.D. and Feldmann, K.A. 1992. Normal and abnormal development in the *Arabidopsis* vegetative shoot apex. *Plant Cell* 4:631–643.

Medford, J.I., Elmer, J.S. and Klee, H.J. 1991. Molecular cloning and character- ization of genes expressed in shoot apical meristems. *Plant Cell* 3:359–370.

Meeks-Wagner, D.R., Dennis, E.S., Tran Thanh Van, K. and Peacock, W.J. 1989. Tobacco genes expressed during *in vitro* floral initiation and their expression during normal plant development. *Plant Cell* 1:25–35.

Meicenheimer, R.D. 1979. Relationships between shoot growth and changing phyllotaxy of *Ranunculus*. *American Journal of Botany* 66:557–569.

Meicenheimer, R.D. 1981. Changes in *Epilobium* phyllotaxy induced by *N*-1-naphthylphthalamic acid and α-4-chlorophenoxyisobutyric acid. *American Journal of Botany* 68:1139–1154.

Meicenheimer, R.D. 1982. Change in *Epilobium* phyllotaxy during reproductive transition. *American Journal of Botany* 69:1108–1118.

Meinhardt, H. 1982. *Models of Biological Pattern Formation*. London: Academic Press.

Meins, F. 1983. Heritable variation in plant cell culture. *Annual Review of Plant Physiology* 34:327–346.

Meins, F. and Lutz, J. 1980. The induction of cytokinin habituation in primary pith explants of tobacco. *Planta* 149:402–407.

Meissner, R. and Theres, K. 1995. Isolation and characterisation of the tomato homeobox gene *THOM1*. *Planta* 195:541–547.

Melzer, S., Majewski, D.M. and Apel, K. 1990. Early changes in gene expression during the transition from vegetative to generative growth in the long-day plant *Sinapis alba*. *Plant Cell* 2:953–961.

Menzel, G., Apel, K. and Melzer, S. 1996. Identification of two MADS box genes that are expressed in the apical meristem of the long-day plant *Sinapis alba* in transition to flowering. *Plant Journal* 9:399–408.

Mestre, J.-C. and Guignard, J.-L. 1967. La mise en place des méristèmes cauli- naire et radiculaire au cours de l'embryogenèse du *Cerastium pumilum* Curt. *Bulletin de la Société Botanique de France* 114:387–396.

Meyer, V.G. 1966. Flower abnormalities. *Botanical Review* 32:165–218.

Meyerowitz, E.M., Bowman, J.L., Brockman, L.L., Drews, G.N., Jack, T., Sieburth, L.E. and Weigel, D. 1991. A genetic and molecular model for flower development in *Arabidopsis thaliana*. *Development, Supplement* 1: 157–167.

Mia, A.J. and Pathak, S.M. 1968. A histochemical study of the shoot apical meristem of *Rauwolfia* with reference to the differentiation of sclereids. *Canadian Journal of Botany* 46:115–120.

Michaux, N. 1968. Etude cytologique du méristème apical du *Pteris cretica* (L.). *C.R. Acad. Sci. Paris D* 267:1442–1444.

Michaux, N. 1969. Durée des phases du cycle mitotique dans le méristème apical de l'*Isoetes setacea* Lam. *C.R. Acad. Sci. Paris D* 269:1396–1399.

Michaux, N. 1970. Détermination, par cytophotométrie, de la quantité d'ADN contenue dans le noyau de la cellule apicale des méristèmes jeunes et adultes du *Pteris cretica* L. *C.R. Acad. Sci. Paris D* 271:656–659.

Michaux, N. 1971. Durée du cycle mitotique dans le méristème apical du jeune sporophyte du *Pteris cretica* L. *C.R. Acad. Sci. Paris D* 273:336–339.

Michaux-Ferrière, N. 1973. Culture et comportement *in vitro* du méristème apical adulte du *Pteris cretica* L. *C.R. Acad. Sci. Paris D* 277:2149–2152.

Michaux-Ferrière, N. 1975. Mise en place de cellules apicales sur un cal du *Pteris cretica* L. cultivé *in vitro*. Etude de l'acquisition de leurs caractères originaux. *C.R. Acad. Sci. Paris, D* 281:783–786.

Michaux-Ferrière, N. 1980. Etude structurale, fonctionnelle et métabolique du méristème apical jeune de *l'Isoetes setacea*. *Canadian Journal of Botany* 58:2506–2512.

Michaux-Ferrière, N. 1981a. Variation de la durée des cycles cellulaires au cours du passage de l'état jeune a l'état adulte dans le méristème caulinaire du *Polypodium vulgare*. *Canadian Journal of Botany* 59:1811–1816.

Michaux-Ferrière, N. 1981b. Quantitative study of RNA in the shoot apex of *Pteris cretica* during its development. *Zeitschrift für Pflanzenphysiologie* 101:233–248.

Miginiac, E. 1972. Cinetique d'action comparee des racines et de la kinetine sur le developpement floral de bourgeons cotyledonaires chez le *Scrofularia arguta* Sol. *Physiologie Végétale* 10:627–636.

Miller, D.R. and Goodin, J.R. 1976. Cellular growth rates of juvenile and adult *Hedera helix* L. *Plant Science Letters* 7:397–401.

Miller, M.B. 1976. The transition from vegetative to floral development in the shoot apex. PhD thesis, University of Edinburgh.

Miller, M.B. and Lyndon, R.F. 1975. The cell cycle in vegetative and floral shoot meristems measured by a double labelling technique. *Planta* 126:37–43.

Miller, M.B. and Lyndon, R.F. 1976. Rates of growth and cell division in the shoot apex of *Silene* during the transition to flowering. *Journal of Experimental Botany* 27:1142–1153.

Miller, M.B. and Lyndon, R.F. 1977. Changes in RNA levels in the shoot apex of *Silene* during the transition to flowering. *Planta* 136:167–172.

Milyaeva, E.L. and Chailakhyan, M.Kh. 1981. Diurnal fluctuation of mitotic activity in stem apices of *Rudbeckia bicolor* during flower evocation. *Physiologia Rastenii* 28:302–306. [In Russian with English abstract.]

Mitchison, G.J. 1977. Phyllotaxis and the Fibonacci series. *Science* 196:271–275.

Molder, M. and Owens, J.N. 1972. Ontogeny and histochemistry of the vegetative apex of *Cosmos bipinnatus* 'Sensation'. *Canadian Journal of Botany* 50:1171–1184.

Molder, M. and Owens, J.N. 1973. Ontogeny and histochemistry of the intermediate and reproductive apices of *Cosmos bipinnatus* var. Sensation in response to gibberellin A3 and photoperiod. *Canadian Journal of Botany* 51:535–551.

Nadeau, J.A., Zhang, X.S., Li, J. and O'Neill, S.D. 1996. Ovule development: identification of stage-specific and tissue-specific cDNAs. *Plant Cell* 8:213–239.

Neilson-Jones, W. 1969. *Plant Chimeras* (2nd Edition). London: Methuen.

Nelson, A.J. 1990a. Net alignment of cellulose in the periclinal walls of the shoot apex surface cells in *Kalanchoë blossfeldiana*. I. Transition from vegetative to reproductive morphogenesis. *Canadian Journal of Botany* 68:2668–2677.

Nelson, A.J. 1990b. Net alignment of cellulose in the periclinal walls of the shoot apex surface cells in *Kalanchoë blossfeldiana*. II. Flower development. *Canadian Journal of Botany* 68:2678–2684.

Nelson, T. and Langdale, J.A. 1989. Patterns of leaf development in C4 plants. *Plant Cell* 1:3–13.

Newman, I. V. 1956. Pattern in meristems of vascular plants. I. Cell partition in living apices and in the cambial zones in relation to the concepts of initial cells and apical cells. *Phytomorphology* 6:1–19.

Niklas, K.J. 1988. The role of phyllotactic pattern as a 'developmental constraint' on the interception of light by leaf surfaces. *Evolution* 42:1–16.

Niklas, K.J. and Mauseth, J.D. 1980. Simulations of cell dimensions in shoot apical meristems: implications concerning zonate apices. *American Journal of Botany* 67:715–732.

Nishino, E. 1983. Corolla tube formation in the Primulaceae and Ericales. *Botanical Magazine Tokyo* 96:319–342.

Nougarède, A. 1967. Experimental cytology of the shoot apical cells during vegetative growth and flowering. *International Review of Cytology* 21:203–351.

Nougarède, A., Di Michele, M.N., Rondet, P. and Saint-Côme, R. 1990. Plastochrone, cycle cellulaire et teneurs en ADN nucléaire du méristème caulinaire de plants de *Chrysanthemum* segetum soumis à deux conditions lumineuses différentes, sous une photopériode de 16 heures. *Canadian Journal of Botany* 68:2389–2397.

Nougarède, A., Francis, D. and Rondet, P. 1991. Location of cell cycle changes in relation to morphological changes in the shoot apex of *Silene coeli-rosa* immediately before sepal initiation. *Protoplasma* 165:1–10.

Nougarède, A., Gifford, E.M. and Rondet, P. 1965. Cytohistological studies of the apical meristem of *Amaranthus retroflexus* under various photoperiodic regimes. *Botanical Gazette* 126:248–298.

Nougarède, A. and Rembur, J. 1976. Le RNA cytoplasmique dans le point végétatif du *Chrysanthemum segetum* L.: concentration, teneur par cellule et synthèse. *Annales des Sciences Naturelles, Botanique, Paris* 17:159–186.

Nougarède, A. and Rembur, J. 1977. Determination of cell cycle and DNA synthesis duration in the shoot apex of *Chrysanthemum segetum* L. by double-labelling autoradiographic techniques. *Zeitschrift für Pflanzenphysiologie* 85:283–295.

Nougarède, A. and Rembur, J. 1978. Variations of the cell cycle phases in the shoot apex of *Chrysanthemum segetum* L. *Zeitschrift für Pflanzenphysiologie* 90:379–389.

Nougarède, A., Rembur, J. and Saint-Côme, R. 1987. Rates of cell division in the young prefloral shoot apex of *Chrysanthemum segetum* L. *Protoplasma* 138:156–160.

Nougarède, A. and Rondet, P. 1973. Un modèle original d'organisation de la tige: étude du fonctionnement plastochronique chez le *Pisum sativum* L. var. nain hâtif d'Annonay. *C.R. Acad. Sci. Paris D* 277:997–1000.

Nougarède, A. and Rondet, P. 1975. Index mitotiques et teneurs en DNA nucléaire du méristème axillaire de la feuille de rang 6 et point végétatif terminal, chez le *Pisum sativum* L. var. nain hâtif d'Annonay. *C.R. Acad. Sci. Paris D* 280:709–712.

Nougarède, A. and Rondet, P. 1976. Durée des cycles cellulaires du méristème terminal et des méristèmes axillaires du *Pisum sativum* L. *C.R. Acad. Sci. Paris D* 282:715–718.

Ogura, Y. 1972. *Comparative Anatomy of Vegetative Organs of the Pteridophytes. Handbuch der Pflanzenanatomie, 8/3*. Berlin: Gebrüder Borntraeger.

Okada, K., Ueda, J., Komaki, M.K., Bell, C.J. and Shimura, Y. 1991. Requirement of the auxin polar transport system in early stages of *Arabidopsis* floral bud formation. *Plant Cell* 3:667–684.

Ono, M., Okazaki, M., Harada, H. and Uchimiya, H. 1988. *In vitro* translated polypeptides of different organs of *Pharbitis nil* Chois, strain Violet, under flower-inductive and non-inductive conditions. *Plant Science* 58:1–7.

Opatrná, J. 1970. Histochemical localization of some dehydrogenases in the shoot meristem of wheat at different stages of development. In *Cellular and Molecular Aspects of Floral Induction*, ed. G. Bernier, pp. 80–90. London: Longman.

Opatrná, J. 1974. Selective histochemical localization of alcohol dehydrogenase in bud primordia cells of wheat shoot apices. *Biologia Plantarum* 16:149–151.

Opatrná, J. 1975. Histochemical investigation of dehydrogenases in shoot apices of wheat plants at different ontogenetic stages. *Biologia Plantarum* 17:67–74.

Opatrná, J., Horavka, B., Ullmann, J. and Krekule, J. 1982. The inhibition and stimulation of DNA synthesis in shoot apices of *Chenopodium rubrum* L. during photoperiodic induction of flowering. *Biologia Plantarum* 24:63–71.

Ormrod, J.C. and Bernier, G. 1990. Cell cycle patterns in the shoot apex of *Lolium temulentum* L. cv. Ceres during the transition to flowering following a single long day. *Journal of Experimental Botany* 41:211–216.

Ormrod, J.C. and Francis, D. 1985. Effects of light on the cell cycle in the shoot apex of *Silene coeli-rosa* L. on the first day of floral induction. *Protoplasma* 124:96–105.

Ormrod, J.C. and Francis, D. 1986a. Cell cycle responses to red or far-red light, or darkness, in the shoot apex of *Silene coeli-rosa* L. during floral induction. *Annals of Botany* 57:91–100.

Ormrod, J.C. and Francis, D. 1986b. Mean rate of DNA replication and replicon size in the shoot apex of *Silene coeli-rosa* L. during the initial 120 minutes of the first day of floral induction. *Protoplasma* 130:206–210.

Ormrod, J.C. and Francis, D. 1987. Effects of interpolated dark periods during the first long day of floral induction on the cell cycle in the shoot apex of *Silene coeli-rosa*. *Physiologia Plantarum* 71:372–378.

Ormrod, J.C. and Francis, D. (eds) 1993. *Molecular and Cell Biology of the Plant Cell Cycle*. Dordrecht: Kluwer.

Orr, A.R. 1978. Inflorescence development in *Brassica campestris* L. *American Journal of Botany* 65:466–470.

Orr, A.R. 1984. Histochemical study of enzyme activity in the shoot apical meristem of *Brassica campestris* L. during transition to flowering. II. Cytochrome oxidase. *Botanical Gazette* 145:308–311.

Orr, A.R. 1985. Histochemical study of enzyme activity in the shoot apical meristem of *Brassica campestris* L. during transition to flowering. III. Glucose-6-phosphate dehydrogenase and 6-phosphogluconate dehydrogenase. *Botanical Gazette* 146:477–482.

Orr, A.R. 1987. Changes in glyceraldehyde 3-phosphate dehydrogenase activity in shoot apical meristems of *Brassica campestris* during transition to flowering. *American Journal of Botany* 74:1161–1166.

Paolillo, D.J. 1984. Cell and plastid cycles. In *The Experimental Biology of Bryophytes*, eds A.F. Dyer and J.G. Duckett, pp. 117–142. London: Academic Press.

Perbal, M.C., Haughn, G., Saedler, H. and Schwarzsommer, Z. 1996. Non-cell autonomous function of the *Antirrhinum* floral homeotic proteins DEFICIENS and GLOBOSA is exerted by their polar cell-to-cell trafficking. *Development* 122:3433–3441.

Pereira, M.F.A. and Dale, J.E. 1982. Effects of 2,4-dichlorophenoxyacetic acid on shoot apex development of *Phaseolus vulgaris*. *Zeitschrift für Pflanzenphysiologie* 107:169–177.

Petersen, K. and Orr, A.R. 1983. Histochemical study of enzyme activity in the shoot apical meristem of *Brassica campestris* L. during transition to flowering. I. Succinic dehydrogenase. *Botanical Gazette* 144:338–341.

Philips, D.J. and Matthews, G.J. 1964. Growth and development of carnation shoot tips *in vitro*. *Botanical Gazette* 125:7–12.

Philipson, W.R. 1954. Organization of the shoot apex in Dicotyledons. *Phytomorphology* 4:70–75.

Pierard, D., Jacqmard, A. and Bernier, G. 1977. Changes in the protein composition of the shoot apical bud of *Sinapis alba* in transition to flowering. *Physiologia Plantarum* 41:254–258.

Pierard, D., Jacqmard, A. and Bernier, G. 1979. Changements de la composition en protéines des différentes parties du bourgeon apical de *Sinapis alba* au cours de sa mise à fleurs. *C.R. Acad. Sci. Paris D* 289:761–763.

Pierard, D., Jacqmard, A., Bernier, G. and Salmon, J. 1980. Appearance and disappearance of proteins in the shoot apical bud of *Sinapis alba* in transition to flowering. *Planta* 150:397–405.

Pillai, K. 1963. Structural and seasonal activity of the shoot apex of some *Cupressus* species. *New Phytologist* 62:335–340.

Pnueli, L., Abu-Abeid, M., Zamir, D., Nacken, W., Schwarz-Sommer, Z. and Lifschitz, E. 1991. The MADS box gene family in tomato: temporal expression during floral development, conserved secondary structures and homology with homeotic genes from *Antirrhinum* and *Arabidopsis*. *Plant Journal* 1:255–266.

Poethig, R.S. 1984. Patterns and problems in angiosperm leaf morphogenesis. In *Pattern Formation: A Primer in Developmental Biology*, ed. G.M. Malacinski, pp. 413–432. New York: Macmillan.

Poethig, R.S. 1987. Clonal analysis of cell lineage patterns in plant development. *American Journal of Botany* 74:581–594.

Poethig, R.S., Coe, E.H. and Johri, M.M. 1986. Cell lineage patterns in maize embryogenesis: a clonal analysis. *Developmental Biology* 117:392–404.

Poethig, R.S. and Sussex, I.M. 1985a. The developmental morphology and growth dynamics of the tobacco leaf. *Planta* 165:158–169.

Poethig, R.S. and Sussex, I.M. 1985b. The cellular parameters of leaf development in tobacco: a clonal analysis. *Planta* 165:170–184.

Polito, V.S. 1979. Cell division kinetics in the shoot apical meristem of *Ceratopteris thalictroides* Brong. with special reference to the apical cell. *American Journal of Botany* 66:485–493.

Polito, V.S. 1980. DNA microspectrophotometry of shoot apical meristem cell populations in *Ceratopteris thalictroides* (Filicales). *American Journal of Botany* 67:274–277.

Polito, V.S. and Alliata, V. 1981. Growth of calluses derved from shoot apical meristems of adult and juvenile English ivy (*Hedera helix* L.). *Plant Science Letters* 22:387–393.

Pomar, M.C., Slabnik, E., Caso, O.H. and Diaz, H. 1986. Leaf dimorphism in *Taraxacum officinale* during *in vitro* culture of shoot tips. *Journal of Plant Physiology* 122:413–421.

Pri-Hadash, A., Hareven, D. and Lifschitz, E. 1992. A meristem-related gene from tomato encodes a dUTPase: analysis of expression in vegetative and floral meristems. *Plant Cell* 4:149–159.

Pritchard, H.B. 1977. Studies on the growth and development of the excised embryo of *Elaeis guineensis* in culture. PhD thesis, University of Edinburgh.

Pritchard, H.N. 1964. A cytochemical study of embryo development in *Stellaria media*. *American Journal of Botany* 51:472–479.

Raju, M.V.S. 1968. Developmental studies on leafy spurge (*Euphorbia esula*). Apices of seedling and adventitious shoots. *Canadian Journal of Botany* 46:1529–1532.

Raju, M.V.S. and Ho, T.W.M. 1973. Developmental studies on leafy spurge (*Euphorbia esula*). Histochemical and autoradiographic studies of the adventitious shoot apices. *Canadian Journal of Botany* 51:211–219.

Rastogi, R. and Sawhney, V.K. 1990a. Polyamines and flower development in the male sterile stamenless-2 mutant of tomato (*Lycopersicon esculentum* Mill.). I. Level of polyamines and their biosynthesis in normal and mutant flowers. *Plant Physiology* 93:439–445.

Rastogi, R. and Sawhney, V.K. 1990b. Polyamines and flower development in the male sterile stamenless-2 mutant of tomato (*Lycopersicon esculentum* Mill.). II. Effects of polyamines and their biosynthetic inhibitors on the development of normal and mutant flower buds cultured *in vitro*. *Plant Physiology* 93:446–452.

Rees, A.R. 1964. The apical organization and phyllotaxis of the oil palm. *Annals of Botany* 28:57–69.

Reeve, R.M. 1948. The 'Tunica–Corpus' concept and development of shoot apices in certain dicotyledons. *American Journal of Botany* 35:65–75.

Rembur, J. 1970. Etude autoradiographique et cytophotométrique de la synthèse du DNA au cours des premières heures de la germination chez le *Xanthium pennsylvanicum* Wallr. (Ambrosiacées). *C.R. Acad. Sci. Paris D* 271:908–911.

Rembur, J. and Nougarède, A. 1977. Duration of cell cycles in the shoot apex of *Chrysanthemum segetum* L. *Zeitschrift für Pflanzenphysiologie* 81:173–179.

Rembur, J. and Nougarède, A. 1987. Microspectrophotometric measurements of DNA by automated computerized scanning in cycling cells of the *Chrysanthemum segetum* shoot apex. *Protoplasma* 136:183–190.

Rembur, J. and Nougarède, A. 1989. Changes in the polypeptide composition during the ontogenic development of the shoot apex of *Chrysanthemum segetum* L. analyzed by two-dimensional mini-gel electrophoresis. *Plant and Cell Physiology* 30:359–363.

Richards, F.J. 1948. The geometry of phyllotaxis and its origin. *Symposia of the Society for Experimental Biology* 2:217–245.

Richards, F.J. 1951. Phyllotaxis: its quantitative expression and relation to growth in the apex. *Philosophical Transactions of the Royal Society of London B* 235:509–564.

Riechmann, J.L., Krizek, B.A. and Meyerowitz, E.M. 1996. Dimerization specificity of *Arabidopsis* MADS domain homeotic proteins APETALA1, APETALA3, PISTILLATA, and AGAMOUS. *Proceedings of the National Academy of Sciences, USA* 93:4793–4798.

Rijven, A.H.G.C. 1968. Randomness in the generation of phyllotaxis. I. The initiation of the first leaf in some Trifolieae. *New Phytologist* 67:247–256.

Rijven, A.H.G.C. 1969. Randomness in the generation of phyllotaxis. II. Initiation of the third leaf in *Trigonella foenum-graecum* L. *New Phytologist* 68:377–386.

Robbins, W.J. and Hervey, A. 1970. Tissue culture of callus from seedling and adult stages of *Hedera helix*. *American Journal of Botany* 57:452–457.

Rogan, P.G. and Smith, D.L. 1975. The effect of temperature and nitrogen level on the morphology of *Agropyron repens* (L.) Beauv. *Weed Research* 15:93–99.

Rogler, C.E. and Hackett, W.P. 1975. Phase change in *Hedera helix*: induction of the mature to juvenile phase change by gibberellin A_3. *Physiologia Plantarum* 34:141–147.

Rolinson, A.E. 1976. Rates of cell division in the vegetative shoot apex of rice (*Oryza sativa* L.). *Annals of Botany* 40:939–945.

Rolinson, A.E. and Vince-Prue, D. 1976. Responses of the rice shoot apex to irradiation with red and far-red light. *Planta* 132:215–220.

Romberger, J.A. and Gregory, R.A. 1977. The shoot apical ontogeny of the *Picea abies* seedling. III. Some age-related aspects of morphogenesis. *American Journal of Botany* 64:622–630.

Running, M.P. and Meyerowitz, E.M. 1996. Mutations in the *PERIANTHA* gene of *Arabidopsis* specifically alter floral organ number and initiation pattern. *Development* 122:1261–1269.

Sachs, T. 1969. Regeneration experiments on the determination of the form of leaves. *Israel Journal of Botany* 18:21–30.

Sachs, T. 1981. The control of patterned differentiation of vascular tissues. *Advances in Botanical Research* 9:151–262.

Sachs, T. 1991. *Pattern Formation in Plant Tissues*. Cambridge: Cambridge University Press.

Saint-Côme, R. 1971. Durée du cycle mitotique chez le *Coleus blumei* Benth. durant les phases préflorale et reproductrice. *C.R. Acad. Sci. Paris D* 272:44–47.

Sakaguchi, S., Hogetsu, T. and Hara, N. 1988. Arrangement of cortical microtubules in the shoot apex of *Vinca major* L. *Planta* 175:403–411.

Sakaguchi, S., Hogetsu, T. and Hara, N. 1990. Specific arrangements of cortical microtubules are correlated with the architecture of meristems in shoot apices of angiosperms and gymnosperms. *Botanical Magazine Tokyo* 103:143–163.

Santiago, J.F. and Goodwin, P.B. 1988. Restricted cell/cell communication in the shoot apex of *Silene coeli-rosa* during the transition to flowering is associated with a high mitotic index rather than with evocation. *Protoplasma* 146:52–60.

Satina, S., Blakeslee, A.F. and Avery, A.G. 1940. Demonstration of the three germ layers in the shoot apex of *Datura* by means of induced polyploidy in periclinal chimeras. *American Journal of Botany* 27:895–905.

Sattler, R. 1973. *Organogenesis of Flowers: A Photographic Text-Atlas*. Toronto: University of Toronto Press.

Sawhney, V.K., Chen, K. and Sussex, I.M. 1985. Soluble proteins of the mature floral organs of tomato (*Lycopersicon esculentum* Mill.). *Journal of Plant Physiology* 121:265–271.

Sawhney, V.K. and Greyson, R.I. 1973a. Morphogenesis of the stamenless-2 mutant in tomato. I. Comparative description of the flowers and ontogeny of stamens in the normal and mutant plants. *American Journal of Botany* 60:514–523.

Sawhney, V.K. and Greyson, R.I. 1973b. Morphogenesis of the stamenless-2 mutant in tomato. II. Modification of sex organs in the mutant and normal flowers by plant hormones. *Canadian Journal of Botany* 51:2473–2479.

Sawhney, V.K., Rennie, P.J. and Steeves, T.A. 1981. The ultrastructure of the central zone cells of the shoot apex of *Helianthus annuus*. *Canadian Journal of Botany* 59:2009–2015.

Schwabe, W.W. 1971. Chemical modification of phyllotaxis and its implications. In *Control Mechanisms of Growth and Differentiation* (Symposia of the Society for Experimental Biology 25), eds D.D. Davies and M. Balls, pp. 301–322. Cambridge: Cambridge University Press.

Schwabe, W.W. 1984. Phyllotaxis. In *Positional Controls in Plant Development*, eds P.W. Barlow and D.J. Carr, pp. 403–440. Cambridge: Cambridge University Press.

Schwabe, W.W. 1998. The role and importance of vertical spacing at the plant apex in determining the phyllotactic pattern. In *Symmetry in Plants*, ed. R.V. Jean. Singapore: World Scientific Press. [in press]

Schwabe, W.W. and Al-Doori, A.H. 1973. Analysis of a juvenile-like condition affecting flowering in the black currant (*Ribes nigrum*). *Journal of Experimental Botany* 24:969–981.

Schwarz-Sommer, Z., Hue, I., Huijser, P., Flor, P.J., Hansen, R., Tetens, F., Lönnig, W.-E., Saedler, H. and Sommer, H. 1992. Characterization of the *Antirrhinum* floral homeotic MADS-box gene deficiens: evidence for DNA binding and autoregulation of its persistent expression throughout flower development. *EMBO Journal* 11:251–263.

Seidlová, F. 1976. Growth correlations and RNA synthesis in different parts of the shoot apical meristem of *Chenopodium rubrum* L. induced to flowering. *Biologia Plantarum* 18:19–25.

Seidlová, F. 1977. Localization of starch in shoot apices of vegetative and photo-periodically induced plants of *Chenopodium rubrum*. *Biologia. Plantarum* 19:387–390.

Seidlová, F. 1980. Sequential steps of transition to flowering in *Chenopodium rubrum* L. *Physiologie Végétale* 18:477–487.

Seidlová, F. and Sádlíková, H. 1983. Floral transition as a sequence of growth changes in different components of the shoot apical meristem of *Chenopodium rubrum*. *Biologia Plantarum* 25:50–62.

Seilhean, V. and Michaux-Ferrière, N. 1985. Cell cycle duration in the meristem of *Nephrolepsis biserrata* stolons: the role of the apical cell. *American Journal of Botany* 72:1089–1094.

Selker, J.M.L. and Green, P.B. 1984. Organogenesis in *Graptopetalum paraguayense* E. Walther: shifts in orientation of cortical microtubule arrays are associated with periclinal divisions. *Planta* 160:289–297.

Selker, J.M.L. and Lyndon, R.F. 1996. Leaf initiation and de novo pattern formation in the absence of an apical meristem and pre-existing patterned leaves in watercress (*Nasturtium officinale*) axillary explants. *Canadian Journal of Botany* 74:625–641.

Selker, J.M.L., Steucek, G.L. and Green, P.B. 1992. Biophysical mechanisms for morphogenetic progressions at the shoot apex. *Developmental Biology* 153:29–43.

Shabde, M. and Murashige, T. 1977. Hormonal requirements of excised *Dianthus caryophyllus* L. shoot apical meristems *in vitro*. *American Journal of Botany* 64:443–448.

Shah, J.J. and Patel, J.D. 1972. The shell zone: its differentiation and probable function in some dicotyledons. *American Journal of Botany* 59:683–690.

Shahar, T., Hennig, N., Gutfinger, T., Hareven, D. and Lifschitz, E. 1992. The tomato 66-3-kD polyphenoloxidase gene: molecular identification and developmental expression. *Plant Cell* 4:135–147.

Shannon, S. and Meeks-Wagner, D. Ry. 1993. Genetic interactions that regulate inflorescence development in *Arabidopsis*. *Plant Cell* 5:639–655.

Sharman, B.C. 1942. Developmental anatomy of the shoot of *Zea mays* L. *Annals of Botany* 6:245–282.

Sharman, B.C. 1947. The biology and developmental morphology of the shoot apex in the Gramineae. *New Phytologist* 46:20–34.

Simon, L. 1973. L'*Impatiens balsamina* L. cultivar 'Buisson Fleuri'. Morphogénèse et ontogénèse expérimentales. Doctoral thesis, University of Nantes.

Simon, R., Carpenter, R., Doyle, S. and Coen, E. 1994. Fimbriata controls flower development by mediating between meristem and organ identity genes. *Cell* 78:99–107.

Sinnott, E.W. 1960. *Plant Morphogenesis*. New York: McGraw-Hill.

Smith, A.G., Hinchee, M. and Horsch, R. 1987. Cell and tissue specific expression localized by *in situ* RNA hybridization in floral tissues. *Plant Molecular Biology Reporter* 5:237–241.

Smith, D.L. and Rogan, P.G. 1979. Growth of the shoot apex of *Agropyron repens* (L.) Beauv. during successive plastochrons. *Annals of Botany* 44:27–34.

Smith, G. 1955. *Cryptogamic Botany. Vol. II. Bryophytes and Pteridophytes*. New York: McGraw-Hill.

Smith, R.H. and Murashige, T. 1970. *In vitro* development of the isolated shoot apical meristem of angiosperms. *American Journal of Botany* 57:562–568.

Smith, R.H. and Murashige, T. 1982. Primordial leaf and phytohormone effects on excised shoot apical meristems of *Coleus blumei* Benth. *American Journal of Botany* 69:1334–1339.

Smyth, D.R., Bowman, J.L. and Meyerowitz, E.M. 1990. Early flower development in *Arabidopsis*. *Plant Cell* 2:755–767.

Snow, M. and Snow, R. 1931. Experiments on phyllotaxis. I. The effect of isolating a primordium. *Philosophical Transactions of the Royal Society of London B* 221:1–43.

Snow, M. and Snow, R. 1933. Experiments on phyllotaxis. II. The effect of displacing a primordium. *Philosophical Transactions of the Royal Society of London B* 222:353–400

Snow, M. and Snow, R. 1937. Auxin and leaf formation. *New Phytologist* 36:1–18.

Snow, M. and Snow, R. 1948. On the determination of leaves. *Symposia of the Society for Experimental Biology* 2:263–275.

Snow, M. and Snow, R. 1952. Minimum areas and leaf determination. *Proceedings of the Royal Society of London B* 139:545–566.

Snow, M. and Snow, R. 1955. Regulation of sizes of leaf primordia by growing-point of stem apex. *Proceedings of the Royal Society of London B* 144:222–229.

Snow, R. 1942. Further experiments on whorled phyllotaxis. *New Phytologist* 41:108–124.

Snow, R. 1965. The causes of the bud eccentricity and the large divergence angles between leaves in Cucurbitaceae. *Philosophical Transactions of the Royal Society B* 250:53–77.

Soetiarto, S.R. and Ball, E. 1969. Ontogenetical and experimental studies of the floral apex of *Portulaca grandiflora*. 2. Bisection of the meristem in successive stages. *Canadian Journal of Botany* 47:1067–1076.

Soma, K. 1958. Morphogenesis in the shoot apex of *Euphorbia lathyris* L. *Journal of the Faculty of Science of the University of Tokyo III* 7:199–256.

Soma, K. 1968. The effect of direct application of 2,4-D to the shoot apex of *Phaseolus vulgaris*. *Phytomorphology* 18:305–324.

Soma, K. and Ball, E. 1963. Studies on the surface growth of the shoot apex of *Lupinus albus*. *Brookhaven Symposia in Biology* 16:13–45.

Soma, K. and Kuriyama, K. 1970. Phyllotactic change in the shoot apex of *Ambrosia artemisiaefolia* var. *elatior* during ontogenesis. *Botanical Magazine Tokyo* 83:13–20.

Sommer, H., Beltran, J.-P., Huitser, P., Pape, H., Lönnig, W.-E., Saedler, H. and Schwarz-Sommer, Z. 1990. *Deficiens*, a homeotic gene involved in the control of flower morphogenesis in *Antirrhinum majus*: the protein shows homology to transcription factors. *EMBO Journal* 9:605–613.

Steeves, T.A. 1966. On the determination of leaf primordia in ferns. In *Trends in Plant Morphogenesis*, ed. E.G. Cutter, pp. 200–219. London: Longman.

Steeves, T.A., Hicks, M.A., Naylor, J.M. and Rennie, P. 1969. Analytical studies on the shoot apex of *Helianthus annuus*. *Canadian Journal of Botany* 47:1367–1375.

Stein, O.L. and Fosket, E.B. 1969. Comparative developmental anatomy of shoots of juvenile and adult *Hedera helix*. *American Journal of Botany* 56:546–551.

Stevens, P.T., Huether, C.A. and Wilson, T.K. 1972. Apical size in the determination of corolla lobe number in *Linanthus androsaecus* ssp. *androsaecus*. *American Journal of Botany* 59:989–992.

Stevenson, D.W. 1976. The cytological and cytohistochemical zonation of the shoot apex of *Botrychium multifidum*. *American Journal of Botany* 63:852–856.

Steward, F.C., Barber, J.T., Bleichert, E.F. and Roca, W.M. 1971. The behaviour of shoot apices of *Tulipa* in relation to floral induction. *Developmental Biology* 25:310–335.

Steward, F.C., Wetmore, R.H. and Pollard, J.K. 1955. The nitrogenous components of the shoot apex of *Adiantum pedatum*. *American Journal of Botany* 42:946–948.

Stewart, R.N. and Dermen, H. 1970. Determination of number and mitotic activity of shoot apical initial cell by analysis of mericlinal chimeras. *American Journal of Botany* 57:816–826.

Stewart, R.N., Meyer, F.G. and Dermen, H. 1972. *Camellia* + 'Daisy Eagleson', a graft chimera of *Camellia sasanqua* and *C. japonica*. *American Journal of Botany* 59:515–524.

Stoutemyer, V.T. and Britt, O.K. 1965. The behavior of tissue cultures from English and Algerian ivy in different growth phases. *American Journal of Botany* 52:805–810.

Sundberg, M.D. 1982. Leaf initiation in *Cyclamen persicum* (Primulaceae). *Canadian Journal of Botany* 60:2231–2234.

Sunderland, N. 1961. Cell division and expansion in the growth of the shoot apex. *Journal of Experimental Botany* 12:446–457.

Sunderland, N. and Brown, R. 1956. Distribution of growth in the apical region of the shoot of *Lupinus albus*. *Journal of Experimental Botany* 7:127–145.

Sunderland, N. and Brown, R. 1976. Development during vegetative growth in the apical region of the shoot of *Lupinus albus*. *Annals of Botany* 40:199–212.

Sunderland, N., Heyes, J.K. and Brown, R. 1957. Protein and respiration in the apical region of the shoot of *Lupinus albus*. *Journal of Experimental Botany* 8:55–70.

Sung, Z.R., Belachew, A., Shuning, B. and Bertrand-Garcia, R. 1992. *EMF*, an *Arabidopsis* gene required for vegetative shoot development. *Science* 258:1645–1647.

Sussex, I.M. 1955. Morphogenesis in *Solanum tuberosum* L.: apical structure and developmental pattern of the juvenile shoot. *Phytomorphology* 5:253–273.

Sussex, I.M. 1964. The permanence of meristems: developmental organizers or reactors to exogenous stimuli? *Brookhaven Symposia in Biology* 16:1–12.

Sussex, I.M. and Rosenthal, D. 1973. Differential ^{3}H-thymidine labeling of nuclei in the shoot apical meristem of *Nicotiana*. *Botanical Gazette* 134:295–301.

Szymkowiak, E.J. and Sussex, I.M. 1992. The internal meristem layer (L3) determines floral meristem size and carpel number in tomato periclinal chimeras. *Plant Cell* 4:1089–1100.

Szymkowiak, E.J. and Sussex, I.M. 1993. Effect of *lateral suppressor* on petal initiation in tomato. *Plant Journal* 4:1–7.

Taylor, M. and Francis, D. 1989. Cell cycle changes in the shoot apex of *Silene coeli-rosa* during the second and third days of floral induction. *Annals of Botany* 64:625–633.

Taylor, M., Francis, D., Rembur, J. and Nougarède, A. 1990. Changes to proteins in the shoot meristem of *Silene coeli-rosa* during the transition to flowering. *Plant and Cell Physiology* 31:1169–1176.

Tepfer, S.S., Nougarède, A. and Rondet, P. 1981. Seasonal studies of vegetative buds of *Helianthus tuberosus*: concentration of nuclei in phase G_1 during winter dormancy. *Canadian Journal of Botany* 59:1918–1927.

Thielke, C. 1965. Strukturwechsel und Enzymmuster am Sprossscheitel einiger Gräser. *Planta* 66:310–319.

Thomas, J.F. and Kanchanapoom, M.L. 1990. Meristematic activity and leaf initiation in the shoot apex of *Nicotiana tabacum* during floral transition. *Botanical Gazette* 151:285–292.

Thompson-Coffe, C., Schoneman Findlay, T.A., Wagner, B.A. and Orr, A.R. 1992. Changes in protein complement accompany early organogenesis in the maize ear. *International Journal of Plant Science* 153:31–39.

Thornley, J.H.M. 1975a. Phyllotaxis. I. A mechanistic model. *Annals of Botany* 39:491–507.

Thornley, J.H.M. 1975b. Phyllotaxis. II. A description in terms of intersecting logarithmic spirals. *Annals of Botany* 39:509–524.

Thornley, J.H.M. and Cockshull, K.E. 1980. A catastrophe model for the switch from vegetative to reproductive growth in the shoot apex. *Annals of Botany* 46:333–341.

Thorpe, T.A. and Murashige, T. 1970. Some histochemical changes underlying shoot initiation in tobacco callus cultures. *Canadian Journal of Botany* 48:277–285.

Tilney-Bassett, R.A.E. 1986. *Plant Chimeras*. London: Arnold.

Tomlinson, P.B. 1961. *The Anatomy of the Monocotyledons. II. Palmae*. Oxford: Oxford University Press.

Torres-Ruiz, R.A. and Jürgens, G. 1994. Mutations in the *FASS* gene uncouple pattern formation and morphogenesis in *Arabidopsis* development. *Development* 120:2967–2978.

Traas, J., Bellini, C., Nacry, P., Kronenberger, K., Bouchez, D. and Caboche, M. 1995. Normal differentiation patterns in plants lacking microtubular prepophase bands. *Nature* 375:676–677.

Tran Thanh Van, K. 1980. Control of morphogenesis by inherent and exogenously applied factors in thin cell layers. *International Review of Cytology, Supplement* 11A:175–194.

Tran Thanh Van, K.M. 1981. Control of morphogenesis in *in vitro* cultures. *Annual Review of Plant Physiology* 32:291–311.

Tröbner, W., Ramirez, L., Motte, P., Hue, I., Huijser, P., Lönnig, W.-E., Saedler, H., Sommer, H. and Schwarz-Sommer, Z. 1992. *GLOBOSA*: a homeotic gene which interacts with *DEFICIENS* in the control of *Antirrhinum* floral organogenesis. *EMBO Journal* 11:4693–4704.

Tucker, S.C. 1984a. Unidirectional organ initiation in leguminous flowers. *American Journal of Botany* 71:1139–1148.

Tucker, S.C. 1984b. Origin of symmetry in flowers. In *Contemporary Problems in Plant Anatomy*, eds R.A. White and W.C. Dickison, pp. 351–394. New York: Academic Press.

Ursin, V.M., Yamaguchi, J. and McCormick, S. 1989. Gametophytic and sporophytic expression of anther-specific genes in developing tomato anthers. *Plant Cell* 1:727–736.

Usciati, M., Codaccioni, M. and Guern, J. 1972. Early cytological and biochemical events induced by a 6-benzylaminopurine application on inhibited axillary buds of *Cicer arietinum* plants. *Journal of Experimental Botany* 23:1009–1020.

Varkey, M. and Nigam, R.K. 1981. Chlorflurenol-induced leafless nodes in *Ricinus communis* L. *Annals of Botany* 47:699–701.

Veen, A.H. and Lindenmayer, A. 1977. Diffusion mechanism for phyllotaxis. Theoretical, physico-chemical and computer study. *Plant Physiology* 60:127–139.

Vince-Prue, D. 1975. *Photoperiodism in Plants*. London: McGraw-Hill.

Wardlaw, C.W. 1949. Further experimental observations on the shoot apex of *Dryopteris aristata* Druce. *Philosophical Transactions of the Royal Society of London B* 233:415–451.

Wardlaw, C.W. 1963. Apical organization and differential growth in ferns. *Journal of the Linnaean Society of London (Botany)* 58:373–385.

Wardlaw, C.W. 1965. The organization of the shoot apex. In *Encyclopedia of Plant Physiology*, 15/1, ed. W. Ruhland, pp. 966–1076. Berlin: Springer.

Weberling, F. 1992. *Morphology of Flowers and Inflorescences*. Cambridge: Cambridge University Press.

Webster, P.L. and MacLeod, R.D. 1980. Characteristics of root apical meristem cell population kinetics: a review of analyses and concepts. *Environmental and Experimental Botany* 20:335–358.

Weigel, D., Alvarez, J., Smyth, D.R., Yanofsky, M.F. and Meyerowitz, E.M. 1992. *LEAFY* controls floral meristem identity in *Arabidopsis*. *Cell* 69:843–859.

Weigel, D. and Meyerowitz, E.M. 1994. The ABCs of floral homeotic genes. *Cell* 78:203–209.

West, W.C. and Gunckel, J.E. 1968a. Histochemical studies of the shoot of *Brachychiton*. I. Cellular growth and insoluble carbohydrates. *Phytomorphology* 18:269–282.

West, W.C. and Gunckel, J.E. 1968b. Histochemical studies of the shoot of *Brachychiton*. II. RNA and protein. *Phytomorphology* 18:283–293.

White, R.A. 1968. A correlation between the apical cell and the heteroblastic leaf series in *Marsilea*. *American Journal of Botany* 55:485–493.

White, R.A. and Turner, M.D. 1995. Anatomy and development of the fern sporophyte. *Botanical Review* 61:281–305.

Wigglesworth, G. 1956. Further notes on *Polytrichum commune* L. *Transactions of the British Bryological Society* 3:115–120.

Williams, M.H. 1991. A sequential study of cell divisions and expansion patterns on a single developing shoot apex of *Vinca major*. *Annals of Botany* 68:541–546.

Williams, R.F. 1975. *The Shoot Apex and Leaf Growth: A Study in Quantitative Biology*. Cambridge: Cambridge University Press.

Wilson, B.F. 1964. A model for cell production by the cambium of conifers. In *The Formation of Wood in Forest Trees*, ed. M.H. Zimmerman, pp. 19–36. New York: Academic Press.

Wimber, D.E. and Quastler, H. 1963. A ^{14}C- and ^{3}H-thymidine double labelling technique in the study of cell proliferation in *Tradescantia* root tips. *Experimental Cell Research* 30:8–22.

Yanofsky, M.F., Ma, H., Bowmann, J.L., Drews, G.N., Feldmann, K.A. and Meyerowitz, E.M. 1990. The protein encoded by the *Arabidopsis* homeotic gene agamous resembles transcription factors. *Nature* 346:35–39.

Zachgo, S., De Andrade Silva, E., Motte, P., Tröbner, W., Saedler, H. and Schwarz-Sommer, Z. 1995. Functional analysis of the *Antirrhinum* floral homeotic *DEFICIENS* gene *in vivo* and *in vitro* using a temperature-sensitive mutant. *Development* 121:2861–2875.

Zagórska-Marek, B. 1994. Phyllotactic diversity in *Magnolia* flowers. *Acta Societatis Botanicorum Poloniae* 63:117–137.

Zeevaart, J.A.D. 1969. *Bryophyllum*. In *The Induction of Flowering: Some Case Histories*, ed. L.T. Evans, pp. 435–456. Melbourne: Macmillan.

Zeevaart, J.A.D. 1976. Physiology of flower formation. *Annual Review of Plant Physiology* 27:321–348.

Zobel, A.M. 1985. The internode of *Sambucus racemosa* L. originates from a single cell layer. *Annals of Botany* 56:105–107.

Zobel, A.M. 1989a. Origin of nodes and internodes in plant shoots. I. Transverse zonation of apical parts of the shoot. *Annals of Botany* 63:201–208.

Zobel, A.M. 1989b. Origin of nodes and internodes in plant shoots. II. Models of node and internode origin from one layer of cells. *Annals of Botany* 63:209–220.

Index

ABC model of flowering gene action, 218–21
abscisic acid, 79
Acacia, 126
acid phosphatase, 105
Acorus, 34
adenylate cyclase, 106
Adiantum, 107
age/ageing, 8, 78, 81, 100, 103, 104, 181, 186
Agropyron, 44, 48, 67, 78, 81, 181
alcohol dehydrogenase, 100
alpine plants, 80
Amaranthus, 98
Ambrosia, 182
amino acid, 107
Anacharis, 120
Anagallis, 18, 59, 120, 221
anisotropic growth, 62, 63
anneau initial, 57
Anomodon, 13
anther, 32, 158, 213, 220, 222, 228, 229
anther-specific genes, 229
anthers, stigmoid, 222
anthocyanin, 20, 229
antibody, 206, 207, 227
anticlinal, *see* cell division plane
antigen, 205–7, 227
Antirrhinum, 112, 145, 151, 187, 212, 214, 216, 217, 219, 220, 225, 226, 229, 230
apical angle, 163, 180

apical cell, 1–15, 52, 60–2, 66, 73, 74, 77, 78, 81, 86, 94, 106, 107, 138
  division sequence, 10–12
apical dome
  diameter, 14, 34, 41, 136, 162, 170, 182–5, 187, 199
  isolated, 191, 224
  maximal/minimal area, 3, 4, 34, 99, 175–9, 182
  redefinition each plastochron, 174–80
  shape, 7, 62–7, 163, 164, 192
apical dominance, 201, 211
apical surface, eccentricity, 164, 165
apical volume, 37, 175, 182–7, 190
Aquilegia, 224
Arabidopsis, 23, 40, 41, 110, 111, 113, 133, 147, 160, 168, 170, 212, 214, 216–20, 223, 225, 228
archegonia, 81
ash, *see Fraxinus*
Aster, 98
autonomy of
  floral organs, 230
  shoot meristem, 67, 195, 200, 221
auxin, 68, 133, 134, 163, 165, 169, 190, 196, 212, 223
  analogue, 134, 163, 191, 196
  antagonist, 134, 163, 164, 191
  synthesis, 191, 192